EVOLUTION AND
OF SELEC

Does natural selection act primarily on individual organisms, on groups, on genes, or on whole species? Samir Okasha provides a comprehensive analysis of the debate in evolutionary biology over the levels of selection, focusing on conceptual, philosophical and foundational questions. A systematic framework is developed for thinking about natural selection acting at multiple levels of the biological hierarchy; the framework is then used to help resolve outstanding issues. Considerable attention is paid to the concept of causality as it relates to the levels of selection, in particular the idea that natural selection at one hierarchical level can have effects that 'filter' up or down to other levels. Unlike previous work in this area by philosophers of science, full account is taken of the recent biological literature on 'major evolutionary transitions' and the recent resurgence of interest in multi-level selection theory among biologists. Other biological topics discussed include Price's equation, kin and group selection, the gene's eye view, evolutionary game theory, outlaws and selfish genetic elements, species and clade selection, and the evolution of individuality. Philosophical topics discussed include reductionism and holism, causation and correlation, the nature of hierarchical organization, and realism and pluralism.

Samir Okasha is Professor of Philosophy of Science at the University of Bristol.

Evolution and the Levels of Selection

SAMIR OKASHA

CLARENDON PRESS · OXFORD

OXFORD
UNIVERSITY PRESS

Great Clarendon Street, Oxford OX2 6DP

Oxford University Press is a department of the University of Oxford.
It furthers the University's objective of excellence in research, scholarship,
and education by publishing worldwide in

Oxford New York

Auckland Cape Town Dar es Salaam Hong Kong Karachi
Kuala Lumpur Madrid Melbourne Mexico City Nairobi
New Delhi Shanghai Taipei Toronto

With offices in

Argentina Austria Brazil Chile Czech Republic France Greece
Guatemala Hungary Italy Japan Poland Portugal Singapore
South Korea Switzerland Thailand Turkey Ukraine Vietnam

Oxford is a registered trade mark of Oxford University Press
in the UK and in certain other countries

Published in the United States
by Oxford University Press Inc., New York

© Samir Okasha 2006

The moral rights of the author have been asserted
Database right Oxford University Press (maker)

First published 2006
First published in paperback 2008

All rights reserved. No part of this publication may be reproduced,
stored in a retrieval system, or transmitted, in any form or by any means,
without the prior permission in writing of Oxford University Press,
or as expressly permitted by law, or under terms agreed with the appropriate
reprographics rights organization. Enquiries concerning reproduction
outside the scope of the above should be sent to the Rights Department,
Oxford University Press, at the address above

You must not circulate this book in any other binding or cover
and you must impose the same condition on any acquirer

British Library Cataloguing in Publication Data
Data available

Library of Congress Cataloging in Publication Data
Data available

Typeset by Laserwords Private Limited, Chennai, India
Printed in Great Britain
on acid-free paper by the
MPG Books Group, Bodmin and King's Lynn

ISBN 978-0-19-926797-2; 978-0-19-955671-7 (Pbk.)

Acknowledgements

I am grateful to numerous friends and colleagues for discussion and correspondence, in particular Peter Godfrey-Smith, Elliott Sober, David Sloan Wilson, John Damuth, Kim Sterelny, the late John Maynard Smith, Jim Griesemer, the late Bill Hamilton, Ayelet Shavit, Lisa Lloyd, Ben Kerr, Patrick Forber, Ken Reisman, Eva Jablonka, Dave Chalmers, Rick Michod, Rob Wilson, Tom Henfrey, Rasmus Winther, Brett Calcott, Melinda Fagan, Denis Walsh, Tim Lewens, Len Nunney, Susanna Rinard, Paul Rainey, Alirio Rosales, James Ladyman, Sahotra Sarkar, Alex Rosenberg, Robert Brandon, Denis Roze, Stefan Lindquist, Bill Wimsatt, Bob Crawford, Alexander Bird, and Jonathan Grose.

I am especially grateful to those who sent me comments on all or part of the manuscript: Lisa Lloyd, Tim Lewens, Roberta Millstein, Peter Godfrey-Smith, Rob Wilson, Neven Sesardic, Ben Kerr, Alirio Rosales, Ayelet Shavit, Rick Michod, and Sahotra Sarkar. I am also grateful to audiences at the Universities of Bristol, Leeds, Austin-Texas, Vienna, Duke, North Carolina, the London School of Economics, and the Australian National University.

I owe a special debt to Peter Godfrey-Smith, with whom I have discussed virtually all of the ideas in this book, often at great length, and who provided extremely detailed feedback on each chapter, saving me from many errors. I am also indebted to Kim Sterelny for his encouragement and advice at all stages of the project, to Alex Rosenberg whose lectures initially aroused my interest in philosophy of biology, to Bill Newton-Smith who supervised my doctoral thesis, to Sahotra Sarkar who read the manuscript for OUP, and to Peter Momtchiloff of OUP who supported the project from the outset.

I began writing this book while based at the Instituto de Investigaciones Filosoficas, National University of Mexico, whose hospitality and financial support I gratefully acknowledge. I am also grateful to the University of York, my former employer, for a term of sabbatical, and to the AHRB for a period of matching leave. I am especially grateful to The Leverhulme Trust, whose financial support enabled me to take research leave in the year 2005–6, and to the University of Bristol, my current employer, for granting me leave. The final manuscript was prepared while I was a Visiting Fellow at the Research School of the

Social Sciences, Australian National University, to whom I am grateful for providing a hospitable working environment. I am grateful to the University of Chicago Press for permission to re-use material published in *Philosophy of Science* 70, 2003. Finally and most of all, I am grateful to Havi Carel for all her love and for keeping me smiling when I ran out of words.

Samir Okasha

Bristol
March 2006

Contents—Summary

Introduction	1
1. Natural Selection in the Abstract	10
2. Selection at Multiple Levels: Concepts and Methods	40
3. Causality and Multi-Level Selection	76
4. Philosophical Issues in the Levels-of-Selection Debate	112
5. The Gene's-Eye View and its Discontents	143
6. The Group Selection Controversy	173
7. Species Selection, Clade Selection, and Macroevolution	203
8. Levels of Selection and the Major Evolutionary Transitions	218
Bibliography	241
Index	257

Contents

Introduction   1

1. Natural Selection in the Abstract   10
    Introduction   10
    1.1 Abstract Formulations of Darwinian Principles   13
    1.2 Price's Equation   18
    1.3 Interpretation of Price's Equation   23
    1.4 Statistical versus Causal Decomposition   25
        1.4.1 Random Drift and Causal Decomposition   31
    1.5 Price's Equation and the Lewontin Conditions   34

2. Selection at Multiple Levels: Concepts and Methods   40
    Introduction   40
    2.1 Hierarchical Organization   40
    2.2 Selection at Multiple Levels: Key Concepts   46
        2.2.1 Particle Characters and Collective Characters   48
        2.2.2 Life Cycles   49
        2.2.3 Particle Fitness and Collective Fitness   53
        2.2.4 The Two Types of Multi-Level Selection   56
        2.2.5 Particle Heritability and Collective Heritability   59
    2.3 Price's Equation in a Hierarchical Setting   62
        2.3.1 The Price Approach to MLS1   62
        2.3.2 Applications   66
        2.3.3 Heritability in MLS1 Revisited   71
        2.3.4 The Price Approach to MLS2   74

3. Causality and Multi-Level Selection   76
    Introduction   76
    3.1 Causes, Correlations, and Cross-Level By-Products   76
    3.2 Selection on Correlated Characters   80
    3.3 Cross-Level By-Products in MLS1   84
        3.3.1 Contextual Analysis: Further Remarks   89
    3.4 Contextual Analysis versus Price's Equation   93

3.5	Cross-Level By-Products in MLS2	100
	3.5.1 Particle→Collective By-Products 100	
	3.5.2 Collective→Particle By-Products 107	

4. Philosophical Issues in the Levels-of-Selection Debate — 112

	Introduction	112
4.1	Emergence and Additivity	112
	4.1.1 The Emergent Character Requirement 112	
	4.1.2 Additivity and the Wimsatt/Lloyd Approach 114	
	4.1.3 Emergent Relations and the Damuth–Heisler Approach 119	
4.2	Screening Off and the Levels of Selection	121
4.3	Realism versus Pluralism about the Levels of Selection	125
	4.3.1 Pluralism and Causality 128	
	4.3.2 Pluralism and Hierarchical Organization 130	
	4.3.3 Pluralism and Multiple Representations 133	
4.4	Reductionism	139

5. The Gene's-Eye View and its Discontents — 143

	Introduction	143
5.1	The Origins of Gene's-Eye Thinking	143
5.2	Genic Selection and the Gene's-Eye View: Process versus Perspective	146
5.3	Outlaws and Genetic Conflicts	149
5.4	Price's Equation versus Contextual Analysis Revisited	154
5.5	Bookkeeping and Causality	158
	5.5.1 The Limits of Genic Accounting 158	
	5.5.2 Sober and Lewontin's Heterosis Argument 162	
5.6	Context-Dependence and the Gene's-Eye View	166
5.7	Reductionism and Pluralism Revisited	169

6. The Group Selection Controversy — 173

	Introduction	173
6.1	Origins of the Group Selection Controversy	174
6.2	Group Selection and the MLS1/MLS2 Distinction	178
6.3	Kin Selection, Reciprocal Altruism, and Evolutionary Game Theory	180
6.4	Maynard Smith versus Sober and Wilson on Group Heritability	185

6.5	The Averaging Fallacy	189
6.6	Random versus Assortative Grouping, Strong versus Weak Altruism	192
6.7	Contextual Analysis versus the Neighbour Approach	198

7. Species Selection, Clade Selection, and Macroevolution — 203

	Introduction	203
7.1	Origins of Species Selection	203
7.2	Genuine Species Selection versus 'Causation from Below'	206
7.3	Species versus Avatars: Damuth's Challenge	210
7.4	The Concept of Clade Selection	212

8. Levels of Selection and the Major Evolutionary Transitions — 218

	Introduction	218
8.1	The Transformation of the Levels-of-Selection Question	219
8.2	Genic versus Hierarchical Approaches to the Transitions	225
8.3	MLS1 versus MLS2 in Relation to Evolutionary Transitions	229
8.4	Michod on Fitness Decoupling and the Emergence of Individuality	233
8.5	Concluding Remarks	236

Bibliography — 241
Index — 257

Introduction

This book is about the 'levels-of-selection' question in evolutionary biology. It is not a work of biology, however, but rather of philosophy of science. It examines a cluster of conceptual, foundational, and philosophical problems raised by the debate over levels of selection and related topics in biology. These problems have been extensively discussed over the past forty years, by both biologists and philosophers of science, resulting in a large and somewhat confusing body of literature. As anyone familiar with the literature knows, there exist a multitude of different vocabularies, conceptual schemes, and mathematical models for thinking about the levels of selection, whose interrelations are not always clear, and a host of competing philosophical analyses. The book aims to bring clarity to this situation by developing a systematic framework for addressing the levels question, and using the framework to help resolve outstanding issues.

I was prompted to write this book for two reasons. First, to help bridge the gap that has opened up between the biological and philosophical literatures. With a few notable exceptions, philosophers' discussions of the levels of selection have not used the language, concepts, and formal techniques used by the biologists themselves. As a result, most philosophical discussions have not had much impact in biology; indeed many biologists simply ignore them. Secondly, recent developments within evolutionary biology itself have led to a substantial reorientation of the traditional levels-of-selection debate, which has yet to be fully reflected in the philosophical discussions. I have in mind the growing body of work on 'major evolutionary transitions', and the realization that multi-level selection is crucial to explaining these transitions (Buss 1987; Maynard Smith and Szathmáry 1995; Michod 1999; Queller 2000; Keller (ed.) 1999; Hammerstein (ed.) 2003). The book aims to take full account of these exciting new developments, and to help integrate the biological and philosophical discussions.

The book's focus is on conceptual and philosophical, rather than empirical, issues. Obviously, empirical data is crucial for resolving the levels-of-selection question, as for all scientific questions; but conceptual clarity is a prerequisite too. Unless we can agree on what it means for there to be selection at a given hierarchical level, on what the criteria for individuating 'levels' are, on whether selection at one level can ever be 'reduced' to selection at another, on how multi-level selection should be modelled, and on whether there is always 'one true fact' about the level(s) at which selection is acting, then there is little prospect of empirical resolution, however much data we collect. Focusing on conceptual questions such as these is not meant to downplay the significance of empirical data, but rather to help provide the clarification needed for addressing the issues empirically.

This conception of the role of philosophy of science—clarifying scientific concepts—will strike some philosophers as conservative. It is true that I assume a fairly sharp distinction between empirical and conceptual questions, an unfashionable view in some quarters. But this does not imply that philosophers must be mere passive observers of science. On the contrary, I think that philosophy can make an invaluable contribution to scientific debates, so long as it is suitably informed. In studying the biological literature on the levels of selection, I have repeatedly been struck by the implicit philosophical assumptions that are made at crucial junctures in the argument, for example, about causation, reductionism, and emergent properties. Scrutinizing these assumptions is a vitally important task, and one that falls naturally to the philosopher of science. The reader is referred to Chapter 3, Section 3.1 for a fuller discussion of why the levels-of-selection debate has a philosophical dimension.

The book is aimed at evolutionary biologists, philosophers of science, and interested parties from other disciplines. It presumes a basic familiarity with Darwinian evolution, but I try to introduce every topic from scratch. Jargon, whether biological or philosophical, is avoided as much as possible, and explained where it is used. In places the treatment is slightly more technical than is customary in philosophical discussions, but no more so than is necessary to achieve clarity. Inevitably, different chapters will appeal more to some readers than others, depending on the reader's interests. The book is designed to be read as a whole, but there is an element of modularity. Chapters 1 to 4, in which a general framework is developed for thinking about the levels of selection, stand together as a unit, with extensive cross-referencing. Subsequent chapters are more self-standing, but they do contain sections that refer back to the previous chapters.

In the remainder of this Introduction, I offer a brief synopsis of the central ideas and arguments contained in each chapter. This is intended as a navigation guide for those who intend to read the whole book, and a consumer guide for those who intend to pick and choose.

In Chapter 1, 'Natural Selection in the Abstract', the logic of evolution by natural selection is spelled out, and the origin of the levels-of-selection problem explained. I emphasize the *abstract* nature of the principle of natural selection—any entities that satisfy the requisite conditions will evolve by natural selection, whatever those entities are. This fact, combined with the fact that the biological world is hierarchically structured, that is, smaller biological units are nested within larger ones, implies that selection may operate at more than one level of the biological hierarchy. This possibility lies at the heart of the levels-of-selection debate, and is what motivates the body of ideas known as 'multi-level selection theory'.

Next I introduce *Price's equation*, a key foundational result in evolutionary theory, which plays a pivotal role in the subsequent discussion. Price's equation, named after the American geneticist George Price, provides a simple, general way of describing an evolving population; it subsumes all more specific evolutionary models as special cases. Though the equation is really no more than a mathematical tautology, it is conceptually invaluable, for it lays bare the essential components of evolution by natural selection in a revealing and formally precise way; in particular, it tells us that *character-fitness covariance* is the essence of natural selection. I briefly explore the link between Price's equation and Lewontin's tripartite account of the conditions required for Darwinian evolution.

The significance of Price's equation for the levels-of-selection question is fourfold. First, given its generality, it provides a framework in which selection at any hierarchical level can be described. Secondly, the equation lends itself naturally to a description of *multi-level* selection, as Price himself realized; for it allows the combined effects of two (or more) levels of selection on a given evolutionary change to be represented in a single scheme. Thirdly, the equation has historical significance, for it played an important part in shaping the debate over group selection (cf. Hamilton 1996; Sober and Wilson 1998). Fourthly, the equation provides an ideal framework for addressing philosophical issues. Since many of these issues have to do with causation, I examine whether Price's description of evolutionary dynamics, which is couched in statistical language, ever admits of a causal interpretation.

In Chapter 2, 'Selection at Multiple Levels: Concepts and Methods', an abstract framework is developed for thinking about selection at multiple hierarchical levels. The first step is to consider the nature of hierarchical organization itself. Typically the biological hierarchy is depicted graphically, with smaller units ('particles') nested within larger ones ('collectives'); but it is not always clear what biological relation(s) are binding the particles into the collectives. The idea that *interaction* among the particles is what binds them into a larger unit is discussed. This 'interactionist' conception of the biological hierarchy is plausible, but it is not the whole story; some unequivocal cases of part–whole structure do not fit it.

At a single level, evolution by natural selection requires character differences, associated differences in fitness, and heritability; so multi-level selection presumably requires these features at more than one hierarchical level. This raises the question: what is the relation between the characters, fitnesses, and heritabilities at the different levels? Restricting the analysis to two levels for simplicity, I discuss how character, fitness and heritability at the collective level might relate to these features at the particle level; this is a logical preliminary to understanding multi-level selection.

Next I explore a well-known ambiguity in the concept of multi-level selection identified by Damuth and Heisler (1988). In multi-level selection 1 (MLS1), the particles are the 'focal' units, that is, the units whose demography gets tracked; the collectives in effect constitute part of the particles' environment. In multi-level selection 2 (MLS2), by contrast, both particles and collectives are focal units. This distinction corresponds to a difference in the relation between the fitnesses at each level. In MLS1, the fitness of a collective is defined as the average fitness of the particles within it; in MLS2, collective fitness is defined independently, though it may on occasion be proportional to average particle fitness. Following Damuth and Heisler, I argue that the MLS1/MLS2 distinction is crucial for clarifying the levels-of-selection issue. In the final part of Chapter 2, I show how both sorts of multi-level selection can be described by the Price equation; this permits a number of important points to be made, in particular, that selection at a lower level corresponds to 'transmission bias' at a higher level.

Introduction

Chapter 3, 'Causality and Multi-Level Selection', analyses the causal dimension of multi-level selection theory. Clearly, Darwinian explanations are causal; to attribute a trait's spread in a population to natural selection is to advance a hypothesis about what *caused* it to spread. We know from Price's equation that selection at any level requires character-fitness covariance at that level; but such covariance need not reflect a direct causal influence of the character on fitness—it can arise for many reasons. Biologists sometimes capture this point by distinguishing between 'direct' and 'indirect' selection on a character.

In a multi-level setting, a further complication arises. It is possible that a character-fitness covariance at one hierarchical level may be a side effect, or by-product, of direct selection at a *different* level (higher or lower). For example, direct selection on individuals living in a group-structured population may lead to a character-fitness covariance at the group level, and thus the *appearance* of a selection process acting directly on the groups. I argue that such 'cross-level by-products' lie at the heart of the levels-of-selection problem; they show that Price's equation is not an infallible guide to determining the level(s) of selection. The key question becomes: when is a character-fitness covariance indicative of direct selection at the level in question, and when is it a by-product of selection acting at a different level? Many of the criteria proposed in the literature for how to determine the 'real' level of selection can be understood as attempts to answer this question.

Cross-level by-products can occur in both the upward and downward directions, and need to be analysed differently for MLS1 and MLS2. The bulk of Chapter 3 is devoted to exploring the nature of cross-level by-products, illustrating them graphically using causal graphs, and examining their philosophical implications. I argue that the concept of a cross-level by-product establishes a link between the levels-of-selection question and the broader philosophical literature on causation in the special sciences. The statistical technique known as 'contextual analysis', which can be used to detect cross-level by-products, is examined; I argue that contextual analysis in effect constitutes a rival to the Price approach to multi-level selection. The relative merits of the two approaches are considered.

Chapter 4, 'Philosophical Issues in the Levels-of-Selection Debate', aims to resolve a number of outstanding philosophical debates over the levels of selection. The 'additivity criterion' of Lloyd and Wimsatt is discussed, as is Brandon's 'screening off' criterion, Vrba's 'emergent character' criterion, and Damuth and Heisler's 'emergent relation' criterion for identifying the level(s) at which selection is acting. I use the analysis of the previous chapter to determine whether, and to what extent, these various criteria are theoretically defensible.

Next I consider the issue of *pluralism* about the levels of selection, a major source of philosophical concern. Pluralists say that in certain circumstances, there is no objective fact about the level(s) at which selection is acting; different answers to the question are equally correct. Realists, by contrast, say that there *is* always an objective fact about the level(s) of selection. I argue that realism is the natural default position, and is the implicit assumption of most biologists. Three different arguments for pluralism are examined. The first derives from a non-realist account of causation; the second from the indeterminacy of hierarchical organization; and the third from the existence of mathematically interchangeable descriptions. I argue that a philosophically interesting form of pluralism is defensible only in very specific circumstances.

Finally, the issue of reductionism is examined. Three different concepts of reductionism that have featured in the levels-of-selection debate are identified. The first is the general idea that properties of wholes should be explained in terms of properties of their parts; the second is the idea that lower levels of selection are explanatorily preferable to higher levels; the third is the idea that selection at one hierarchical level may be 'reducible' to selection at a different level. I argue that these three ideas are logically independent of each other.

Chapter 5, 'The Gene's-Eye View and its Discontents', examines the genic view of evolution associated with Williams, Dawkins, Maynard Smith, and others. The origins of the genic approach in the work of Fisher and Hamilton are traced. I then discuss an ambiguity over its status: is it an empirical thesis about the course of evolution, or a heuristic perspective for thinking about evolution? The ambiguity is resolved by distinguishing genic selection, which is a causal process, from the gene's-eye viewpoint, which is a perspective. Genic selection occurs when there is selection between the genes within a single organism, or genome; it is thus a distinct level of selection of its own. By contrast, a gene's-eye view can be adopted on selection processes occurring at various hierarchical levels, not just the genic level.

Next I discuss outlaw genes, also known as selfish genetic elements (SGEs). These genes are favoured by genic selection but typically opposed by selection at higher levels, leading to intra-genomic conflict.

This suggests that multi-level selection may be useful for understanding SGEs, for their evolutionary dynamics involve selection at more than one level. This in turn permits a number of conceptual points about intra-genomic conflict to be made. I briefly revisit an issue from Chapter 3—the tension between the Price and contextual approaches to multi-level selection. Interestingly, the Price approach proves superior for analysing intra-genomic conflict, despite the general theoretical argument in favour of contextual analysis.

Finally, a number of objections to gene's-eye thinking are examined; these include the charge of 'confusing bookkeeping with causality', the charges of reductionism and genetic determinism, and the charge that the context-dependence of genes' effects on phenotype makes it inappropriate to think in terms of selection on single genes. Some of these objections are defused by invoking the distinction between the process of genic selection and the gene's-eye viewpoint; others are partially valid. Sober and Lewontin's well-known heterosis argument is discussed, as is the old question of 'beanbag genetics'. I argue that the heuristic value of the gene's-eye view is greatest when the genotype–phenotype map is relatively simple.

Chapter 6, 'The Group Selection Controversy', examines the notorious issue of group selection in behavioural ecology, one of the mainstays of the traditional levels-of-selection debate. The origins of the controversy and its subsequent development are traced, up to and including the neo-group selectionist revival of recent years. The relationship between group selection, kin selection, and evolutionary game theory is discussed; I examine the argument that the latter two theories constitute versions of, rather than alternatives to, traditional group selection.

Next I consider a dispute between Maynard Smith and Sober and Wilson over the status of 'trait-group' models. Maynard Smith argues that trait groups cannot be 'units of evolution', for they lack 'heredity' so cannot evolve adaptations; Sober and Wilson dispute this argument. Drawing on the analysis of previous chapters, I argue that both parties are partly right. The key to resolving the dispute is to distinguish two concepts of group heritability, and to keep the distinction between MLS1 and MLS2 clearly in focus. I then discuss what Sober and Wilson call the 'averaging fallacy', a way of defining group selection out of existence by averaging fitnesses across groups; I argue that they are correct to identify this as fallacious.

Lastly, I look at the distinction between 'strong' and 'weak' altruism, and some related arguments of L. Nunney about the correct way to define individual and group selection. Strongly altruistic behaviours are ones that involve an absolute reduction in fitness for the donor; weakly altruistic behaviours boost the donor's absolute fitness, but boost that of others in the group by even more. Nunney's thesis that group selection requires non-random assortment of genotypes, and that individual selection should *not* be defined in terms of within-group fitness differences but rather by the 'mutation test', are critically discussed. These issues prove to be related to the discussion of cross-level by-products in Chapter 3.

Chapter 7, 'Species Selection, Clade Selection, and Macroevolution' is a short chapter discussing selection at the level of species and clades. These modes of selection are usually regarded as relatively minor, though Gould (2002) defends their importance. The history of the species selection debate is outlined, including its conceptual link to the idea that species are individuals. Next I discuss the problem of how to distinguish 'real' species selection from what Vrba and others call 'species sorting', that is, differential speciation/extinction that is a side effect of causal processes at lower hierarchical levels. I argue that Vrba overlooks the important distinction between lower-level *selection* being the cause of differential speciation/extinction, and *some lower-level processes or other* being causally responsible.

I then examine an argument of J. Damuth, who holds that whole species are not the right sorts of entity to figure in a selection process, since they are usually not geographically or ecologically localized. Finally, the concept of clade selection is considered. I argue that since clades are by definition monophyletic, they cannot form parent–offspring lineages as a matter of logic. This implies that 'clade reproduction' is impossible; so clades cannot evolve by a process of cumulative selection.

Chapter 8, 'Levels of Selection and the Major Evolutionary Transitions', looks at the major evolutionary transitions, and in particular the idea that multi-level selection theory is crucial for understanding them. As characterized by Michod (1999) and Maynard Smith and Szathmáry (1995), these transitions occur when a number of free-living biological individuals, capable of surviving and reproducing alone, become integrated into a cooperative whole, generating a new level of biological organization. Such transitions have occurred numerous times in the history of life. Clearly, evolutionary transitions create the potential for conflict between levels of selection, for selection between the smaller units may disrupt the well-being of the collective.

I argue that the traditional levels-of-selection question has been subtly transformed by recent work on evolutionary transitions. In traditional discussions, the existence of the biological hierarchy was taken for granted; the question was about selection and adaptation at pre-existing hierarchical levels. But the evolutionary transitions literature is concerned with the origins of hierarchical organization itself; this requires a 'diachronic' rather than a 'synchronic' formulation of the levels-of-selection question. The implications of this change in perspective are examined. I then consider Buss's contrast between 'genic' and 'hierarchical' approaches to studying the transitions; I argue that the two approaches are complementary, not antithetical.

Finally, I ask what becomes of the distinction between MLS1 and MLS2 in relation to the major transitions. Which type of multi-level selection is the relevant one? I argue that both types are relevant, but at different temporal stages of a transition. In the early stages, when the collectives are loose aggregates of interacting particles, MLS1 is relevant; but later in the transition, when the collectives are cohesive units, capable of bearing autonomously defined fitnesses, MLS2 starts to operate. I illustrate this idea with reference to recent work by Michod and co-workers on the evolution of multicellularity.

1
Natural Selection in the Abstract

INTRODUCTION

The levels-of-selection problem is one of the most fundamental in evolutionary biology, for it arises directly from the underlying logic of Darwinism. The problem can be seen as the upshot of three factors, each of which was appreciated to some extent by Darwin himself. The first and most fundamental factor is the *abstract nature* of the principle of natural selection. Darwin argued that if a population of organisms vary in some respect, and if some variants leave more offspring than others, and if parents tend to resemble their offspring, then the composition of the population will change over time—the fittest variants will gradually supplant the less fit. But it is easy to see that Darwin's reasoning applies not just to individual organisms. *Any* entities which vary, reproduce differentially as a result, and beget offspring that are similar to them, could in principle be subject to Darwinian evolution. The basic logic of natural selection is the same whatever the 'entities' in question are.

The second factor is the *hierarchical organization* that characterizes the biological world. The entities biologists study form a nested hierarchy, lower-level ones properly included within higher-level ones. Multicelled organisms, the traditional focus of evolutionary biology, lie somewhere in the middle of the hierarchy. Each organism is composed of organs and tissues, which are themselves made up of cells; each cell contains a number of organelles and a cell nucleus; each nucleus contains a number of chromosomes; and on each chromosome lie a number of genes. Above the level of the organism we find entities such as kin groups, colonies, demes, species, and whole ecosystems. This hierarchical structure is obvious to us today, but it is not a logically necessary feature of the biological world. Moreover, since the earliest life forms were presumably not hierarchically complex, the various levels in the hierarchy must somehow have evolved.

How exactly the biological hierarchy should be described, that is, which levels should be recognized and why, is a substantive issue that we shall return to. But one point is clear from the outset. Entities at various hierarchical levels, above and below that of the organism, can satisfy the conditions required for evolution by natural selection. For just as organisms give rise to other organisms by reproduction, so cells give rise to other cells by cell division, genes to other genes by DNA replication, groups to other groups by fission (among other ways), species to other species by speciation, and so on. Thus the Darwinian concept of fitness, that is, expected number of offspring, applies to entities of each of these types. So in principle, these entities could form populations that evolve by natural selection.

The third factor concerns not the process of natural selection but its product. Natural selection leads organisms to evolve *adaptations*—traits that enhance their chance of survival and reproduction. The existence of organismic adaptations, many of them exquisitely fine-tuned to environmental demands, shows the importance of organism-level selection in shaping the biota. But organisms also exhibit features that do not seem to benefit them individually, so cannot have evolved in this way. Altruistic behaviour, in which one organism performs an action which benefits another at a cost to itself, is an example. Selection at the level of the individual organism should disfavour altruistic behaviour, for altruists suffer a fitness disadvantage relative to their selfish counterparts, yet such behaviour is quite common. One possible explanation, first canvassed by Darwin himself, is that altruism may have evolved by selection at higher levels of organization, for example, group- or colony-level selection. Groups containing a high proportion of altruists might have a selective advantage over groups contain a preponderance of selfish types, even though within each group, selection favours selfishness (Darwin 1871).

The case of altruism illustrates an important principle, namely that what is advantageous at one hierarchical level may be disadvantageous at another level, leading to potential conflict. Various features of modern organisms suggest the importance of such inter-level conflicts. Mammalian cancer is an example. Cancer obviously cannot be interpreted as an organismic adaptation, for it is often fatal to the individual organism; nor is there any obvious advantage to higher-level entities. But cancer in effect involves a process of *cellular* selection, for cancerous cells increase in frequency relative to other cell lineages within the organism's soma. So a maladaptive feature of individual organisms is explained by selection at a lower hierarchical level, in this case the cellular level. Similarly, selection between the different genes within the same genome, or between nuclear and mitochondrial genes, can have effects that are detrimental for the organism as a whole. For example, mitochondrial genes gain an advantage if they can cause their hosts to produce a preponderance of female offspring, for they are only transmitted maternally. Where they succeed, then a trait that is suboptimal for the organism itself—producing a female-biased brood—is again explained by selection acting at a lower hierarchical level.

The levels-of-selection question results from the interaction of the three factors described above. The abstract character of the principle of natural selection, combined with the hierarchical nature of the biological world, implies that selection *can* operate at levels other than that of the individual organism; and the existence of phenomena that defy interpretation in terms of organismic advantage suggest that this *has* actually happened. The basic elements of this picture have been in place for a long time—Weismann (1903) saw clearly that selection could potentially operate at multiple hierarchical levels, as Gould (2002) has emphasized.[1] But it is only recently that its full significance has been appreciated.[2] Multi-level selection theory plays an increasingly prominent role in the evolutionary literature, and has been applied to a diverse range of biological phenomena (Frank 1998; Michod 1997, 1999; Maynard Smith and Szathmáry 1995; Keller (ed.) 1999; Sober and Wilson 1998; Wilson 1997; Hammerstein (ed.) 2003; Gould 2002; Rice 2004).

This chapter sets the stage for the examination of the levels-of-selection question that follows. Section 1.1 examines in more detail the abstract nature of the Darwinian principles. Section 1.2 provides an introduction to Price's equation, a central foundational result in evolutionary theory, which will play an important role in subsequent chapters. Section 1.3 discusses the interpretation of Price's equation, while Section 1.4 asks whether it constitutes a 'causal decomposition'

---

[1] Weismann (1902) wrote that the 'extension of the principle of natural selection to all grades of vital units is the characteristic feature of my theories . . . this idea will endure even if everything else in the book should prove transient,' (quoted in Gould (2002) p. 223).

[2] A comprehensive history of the levels-of-selection debate has yet to be written. Detailed accounts of various aspects of the history are found in Sober and Wilson 1998, Buss 1987, Segerstråle 2000, and Gould 2002.

of evolutionary change. Section 1.5 examines the link between Price's equation and Lewontin's tripartite analysis of the conditions required for Darwinian evolution.

1.1 ABSTRACT FORMULATIONS OF DARWINIAN PRINCIPLES

Though it is widely agreed that the Darwinian principles can be characterized abstractly, without reference to any specific level of biological organization, the literature contains a number of non-equivalent characterizations. For example, some authors distinguish *units* of selection from *levels* of selection; others distinguish units of *selection* from units of *evolution*; still others recognize neither of these distinctions. Some authors argue that evolution by natural selection requires *two* types of entities, replicators and interactors, while others offer analyses in terms of a single type of entity. Still others argue that reproduction, rather than replication, is the fundamental notion. To some extent these are questions of terminological preference, though there are substantive issues at stake too. For clarity in what follows, conceptual and terminological uniformity is required.

In a well-known article, Lewontin (1970) identified three principles that he said 'embody the principle of evolution by natural selection', namely phenotypic variation, differential fitness, and heritability; entities possessing these properties he called 'units of selection' (p. 1). Fitness he defined as rate of survival and reproduction, and heritability as parent–offspring correlation. Oddly, Lewontin required that the differences in *fitness*, rather than the phenotypic differences, should be heritable, that is, the parent–offspring correlation should hold with respect to fitness rather than phenotype. This is odd because if selection is to produce cross-generational phenotypic change, it is the phenotypic differences, not the fitness differences, that must be heritable.[3] But leaving this oddity aside, Lewontin's formulation seems to capture the essence of the Darwinian process very neatly. In a similar vein, Maynard Smith (1987a) wrote that evolution by natural selection will operate on

---

[3] Heritability of fitness is required if selection is to lead mean population fitness to increase over generations, as Fisher's (1930) 'fundamental theorem' states. But if by evolutionary change one means change in mean phenotype, rather than mean fitness, as Lewontin does, then it is the phenotypic differences, not the fitness differences, that must be heritable.

any entities that exhibit 'multiplication, variation and heredity', so long as the variation affects the probability of multiplying. Entities satisfying these three criteria he called 'units of evolution' rather than selection; here I stick with Lewontin's terminology.[4]

Note that both Lewontin and Maynard Smith treat the relation of reproduction or multiplication as primitive; neither offers an account of what it means for one entity to multiply, or to produce an offspring entity. Griesemer (2000) argues that this is a significant lacuna, and offers an analysis of what reproduction amounts to, based on two key ideas. First, there should be 'material overlap' between parent and offspring entities. This means that offspring must contain, as physical parts, objects or structures that used to be physical parts of their parents. Organismic reproduction, cell division, DNA replication, speciation, and 'demic reproduction' all satisfy this criterion, Griesemer argues: in each case, a physical part of the parent becomes a physical part of its offspring. Secondly, the capacity to reproduce is something that entities must *acquire*; they are not born with it. In effect, this second requirement means that entities capable of reproduction must develop, or have a life cycle.

Griesemer's account is a plausible way of fleshing out the abstract notion of reproduction, and forges an interesting link between developmental and evolutionary processes. But in the interests of maximum generality, I prefer to work with a purely abstract notion. However, Griesemer's account does bring out one important feature of the Lewontin–Maynard Smith characterization that can go unnoticed. This is that reproduction, or multiplication, is generally understood to mean the production of offspring entities that occupy *the same level in the biological hierarchy as the parental entity*. Thus when organisms reproduce they give rise to offspring *organisms*, when cells divide they give rise to offspring *cells*, when colonies reproduce they give rise to offspring *colonies*, and so on. This is the most intuitive way to think of reproduction, and unless otherwise stated is what I shall mean by the term. But as we shall see, there are contexts where reproduction at one level has been defined in terms of the production of offspring entities at another level.

---

[4] In one respect Maynard Smith's terminology is superior, given that the three criteria really describe necessary and sufficient conditions for there to be an evolutionary *response* to selection, rather than for selection to occur. However, the label 'unit of evolution' is sometimes used in a quite different sense, as in the title of Ereshefsky's (1992) collection, for example.

Dawkins (1976, 1982) and Hull (1981) offered a somewhat different characterization of the Darwinian process. In Hull's version, evolution by natural selection occurs when 'environmental interaction' leads to 'differential replication'; it thus involves two types of entity—interactors and replicators. Dawkins spoke of 'vehicles' in place of Hull's 'interactors'. Replicators are defined as any entities of which accurate copies are made—they 'pass on their structure intact' from one generation to another and are characterized by their 'copying fidelity' and 'longevity'. Interactors are defined as entities which 'interact as a cohesive whole with their environment' so as to cause the differential transmission of replicators. Dawkins and Hull argued that the expression 'unit of selection', as it appeared in the early literature, was often ambiguous between replicators and interactors, leading to equivocation.

Despite its popularity, there are reasons for doubting that the Dawkins–Hull characterization offers a fully general account of Darwinian evolution, applicable across the board. One is simply that the Lewontin–Maynard Smith characterization does seem fully general, and involves just *one* type of entity, not two. Gould (2002) argues that treating the replicator–interactor account as fundamental leads to a 'historical paradox', given Darwin's own views on inheritance. If blending rather than particulate inheritance had turned out to be correct, then replicators as defined by Dawkins and Hull would not exist, so replication cannot be essential to the Darwinian process, he argues. Gould's 'paradox' is a dramatic way of making a valid point, namely that what matters for evolution by natural selection is sufficient parent–offspring resemblance, or heritability; the transmission of replicating particles from parent to offspring is not in itself necessary (cf. Godfrey-Smith 2000a).

Another way of appreciating this point is to note that cultural and behavioural, as well as genetic, inheritance can generate the parent–offspring similarity needed for an evolutionary response to selection (Avital and Jablonka 2000; Boyd and Richerson 2005). These inheritance channels do not involve particles bequeathing 'structural copies' of themselves to succeeding generations. So evolutionary changes mediated by cultural and behavioural inheritance cannot be described as the differential transmission of replicators.[5] This suggests that the

---

[5] As Avital and Jablonka say, 'the replicator concept is associated with a very specialized type of information transmission, which does not cover all types of inheritance, and therefore cannot be the basis of all evolution.' (2000 p. 359).

replicator–interactor conceptualization is not a fully general account of Darwinian evolution.[6] Therefore, I do not employ the Dawkins–Hull framework in what follows; the theoretical work done by the replicator/interactor distinction can be captured in other ways, permitting us to remain within the simpler Lewontin–Maynard Smith framework.

Griesemer (2000) raises a quite different objection to the generality of the Dawkins–Hull framework, namely that it characterizes the evolutionary process in terms of features that are themselves the product of evolution. The longevity and copying fidelity of replicators (such as genes) and the cohesiveness of interactors (such as organisms) are highly *evolved* properties, themselves the product of many rounds of cumulative selection. The earliest replicators must have had extremely *poor* copying fidelity (Maynard Smith and Szathmáry 1995), and the earliest multicelled organisms must have been highly *non*-cohesive entities, owing to the competition between their constituent cell-lineages (Buss 1987; Michod 1999). If we wish to understand how copying fidelity and cohesiveness evolved in the first place, we cannot build these notions into the very concepts used to describe natural selection.

This is an important consideration, whose implications extend beyond the question of the suitability of the Dawkins–Hull framework. It highlights a subtle transformation in the levels-of-selection question since the discussions of the 1960s and 1970s (Griesemer 2000; Okasha 2006). These early discussions tended to take the existence of the biological hierarchy for granted, as if hierarchical organization were simply an exogenous fact about the living world. But of course the biological hierarchy is *itself* the product of evolution—entities further up the hierarchy, such as eukaryotic cells and multicelled organisms, obviously have not existed since the beginning of life on earth. So ideally, we would like an evolutionary theory which explains how the biological hierarchy came into existence, rather than treating it as a given. From this perspective, the levels-of-selection question is not simply about identifying the hierarchical levels(s) at which selection *now* acts, which is how it was traditionally conceived, but about identifying the mechanisms which led the various hierarchical levels to evolve in the first place. Increasingly, evolutionary theorists have turned their attention to this latter question.

---

[6] Symptomatic of this is the fact that attempts to force all selection processes into the replicator–interactor framework often involve significant departures from the original definitions of 'replicator'. This point is noted by Szathmáry and Maynard Smith (1997) who credit it to J. Griesemer.

This new 'diachronic' perspective gives the levels-of-selection question a renewed sense of urgency. Some biologists were inclined to dismiss the traditional debate as a storm in a teacup—arguing that in practice, selection on individual organisms is the only important selective force in evolution, other theoretical possibilities notwithstanding. But as Michod (1999) stresses, multicelled organisms did not come from nowhere, and a complete evolutionary theory must surely try to explain how they evolved, rather than just taking their existence for granted. So levels of selection other than that of the individual organism must have existed in the past, whether or not they still operate today. From this expanded point of view, the argument that individual selection is 'all that matters in practice' is clearly unsustainable.

It would be a mistake to make *too* much of this change in perspective. Griesemer (2000) argues that the problem of explaining the emergence of new levels is 'conceptually prior' to that of explaining the evolution of adaptations at pre-existing levels (p. 70); similarly, Fontana and Buss (1994) say that 'selection cannot set in until there are entities to select' (p. 761). In a sense this is obviously true. But even so, the two explanatory problems are not wholly disjoint. Michod (1999) has recently argued that groups of lower-level entities only count as new individuals themselves, and thus generate a new level in the hierarchy, when they evolve a special *type* of adaptation, namely policing mechanisms to regulate the selfish tendencies of their members. Prior to this stage the groups are merely loose collections of lower-level entities, not genuine evolutionary individuals. If something like this is correct, then the evolution of new levels in the hierarchy cannot be regarded as entirely prior to the evolution of adaptations at those levels. For what converts the group into a true biological unit is precisely the evolution of a special sort of group-level adaptation (cf. Frank 1995b; Szathmáry and Wolpert 2003).

I return to this issue in detail in Chapter 8, when I look at the application of multi-level selection theory to the 'major transitions' in evolution. But for the moment, the important point is this. Since the levels-of-selection debate now encompasses questions about the origin of the biological hierarchy, not just the evolution of adaptations at pre-existing hierarchical levels, an abstract characterization of Darwinian principles cannot refer to highly evolved features, of either organisms or genetic systems, on pain of an inevitable loss of generality. Characterizations in terms of 'high-fidelity replication' and 'cohesiveness' fall foul of this constraint; arguably, those which describe evolution in terms of 'information transfer' do so too (e.g. Williams 1992; Odling-Smee, Laland, and Feldman 2003).[7] As we shall see later, the same constraint tells against certain conceptions of what is required for selection to act at a given hierarchical level, for example, the 'emergent character' requirement. Such requirements mistake a product of Darwinian evolution for a prerequisite of it. This is a consideration in favour of the abstract Lewontin characterization.

The expressions 'unit of selection' and 'level of selection' have engendered certain confusion. The following convention will be observed here: if entities at hierarchical level X are units of selection in the Lewontin sense, I shall say that selection 'operates at level X'. The level of selection is simply the hierarchical level occupied by the entities that are units of selection. Thus we can translate easily between talks of units and levels. Note that this convention contrasts with the usage of Brandon (1988), who uses the unit/level distinction in lieu of the replicator/interactor distinction. It also contrasts with the usage of Reeve and Keller (1999), who regard the 'units of selection' question as stale but the 'levels of selection' question as empirically exciting. By the former, they mean the 'gene versus organism' debate prompted by Dawkins's work; by the latter, they mean questions about evolutionary transitions of the sort discussed above.

To summarize, I favour the original Lewontin characterization of the Darwinian principles as a starting point. A population of entities evolves by natural selection where heritable differences between the entities lead to differences in their reproductive output; reproduction is understood as giving rise to an offspring entity that occupies the same hierarchical level as the parent, unless otherwise stated. Entities satisfying these conditions are units of selection; the level in the hierarchy which the entities occupy is the level of selection. This characterization has the virtues of simplicity and generality, though certain complications will emerge. In Section 1.5 we shall see how to integrate it with an abstract mathematical description of the evolutionary process.

1.2 PRICE'S EQUATION

Price's equation, first published by George Price (1972), is a simple algebraic result that describes a population's evolution from one generation

---

[7] This is because on the standard accounts of what genetic information is, genes contain information as a result of evolutionary processes (cf. Maynard Smith 2000, Sterelny 2000).

to another. The power of the equation lies in its generality: unlike most formal descriptions of the evolutionary process, it rests on no contingent biological assumptions, so always holds true (cf. Frank 1995a, 1998). Moreover, the equation lays bare the essential components of evolution by natural selection in a highly revealing way.

Price's equation actually has special significance for the levels-of-selection question, for reasons that go beyond its generality.[8] For the equation lends itself very naturally to a description of selection at multiple hierarchical levels, as Price himself realized. This theme was developed by Hamilton (1975) in a well-known paper, but it is only recently that its full significance has become apparent. Grafen (1985) reported that he could only find two papers, other than Hamilton's, that made use of Price's methods at any length;[9] today those methods are very widely used, particularly by theorists interested in multi-level and hierarchical approaches to selection, for example, Frank (1998), Michod (1997, 1999), Queller (1992b), Damuth and Heisler (1988), Tsuji (1995), Sober and Wilson (1998), Rice (2004), and others. I will argue that the Price formalism provides an ideal framework for addressing philosophical questions about the levels of selection.

A simple derivation of the basic Price equation is given below, in relation to a single level of selection. Application of the formalism to multiple levels is postponed until Chapter 2.

Consider a population containing n entities, called the P-population (for parental). It doesn't matter what the entities are. The entities vary with respect to a measurable phenotypic character z, the evolution of which interests us. We let z_i denote the character value of the ith entity, and $\bar{z}$ the average character value in the whole population, i.e. $\bar{z} = \frac{1}{n} \sum_1^n z_i$. So for example, if z were height, then z_i would be the height of the ith entity and $\bar{z}$ the average height of the whole population.

If the character z is selectively significant, we might expect the quantity $\bar{z}$ to change over time. To track this change, we need to take account of fitness. We let w_i denote the absolute fitness of the ith entity, defined as the total number of offspring entities it produces. For simplicity

---

[8] The history of Price's equation, and its implications for the group selection question in particular, are discussed by Hamilton (1996), Frank (1995a), Sober and Wilson (1998), and Segerstråle (2000).

[9] The papers Grafen cites are Seger (1981) and Wade (1985); he could also have mentioned Arnold and Fristrup (1982).

we will assume that reproduction is asexual.[10] Average fitness in the P-population as a whole is $\bar{w} = \frac{1}{n}\sum_1^n w_i$. The relative fitness of the ith entity is therefore $\omega_i = w_i/\bar{w}$.

To track the evolution of $\bar{z}$, we also need to take account of how the character z is transmitted from parent to offspring. We let z'_i denote the average character value of the *offspring* of the ith entity. If transmission is perfect, that is, if each parental entity transmits its character to each of its offspring with no deviation, then $z'_i = z_i$ for each i. However, transmission may not be perfect: offspring may deviate from their parents with respect to z. We define Δz_i as the difference between the character value of the ith entity and the average for its offspring, that is, $\Delta z_i = z'_i - z_i$. So Δz_i measures the *transmission bias* of the ith entity with respect to the character z. The closer that Δz_i is to zero, the more faithfully the ith entity transmits the character.

If the ith entity leaves no offspring, that is, $w_i = 0$, then by convention we let Δz_i equal the transmission bias that *would* have resulted, if it had left offspring. (This convention is innocent, for in the Price equation the term Δz_i appears multiplied by w_i, so if $w_i = 0$, the value of Δz_i can be arbitrarily chosen. The point of the convention will become clear later.) The average transmission bias in the whole population we will denote by $E(\Delta z_i)$, where 'E' stands for expected value. Obviously, $E(\Delta z_i) = \frac{1}{n}\sum_1^n \Delta z_i$.

Now consider another population of entities, called the O-population, which comprises all the offspring of entities from the P-population.[11] We let $\bar{z}_o$ denote the average character value in the O-population. So if evolution has taken place, $\bar{z}_o$ will be different from $\bar{z}$. To calculate $\bar{z}_o$, note that the O-population is in effect made up of n disjoint subpopulations, where each subpopulation contains all the w_i offspring of the ith entity (see Figure 1.1). By definition, the average character value of the ith subpopulation is z'_i. So the average character value in the O-population as a whole is the weighted average of all the z'_i, the weights determined by subpopulation size w_i. Therefore $\bar{z}_o = \frac{1}{n}\sum_1^n \frac{w_i}{\bar{w}} z'_i$.

Realizing that this is a correct formula for $\bar{z}_o$ is the key to understanding the Price equation. As Frank (1998) has stressed, the formula's peculiarity lies in the fact that although $\bar{z}_o$ denotes a property of the

---

[10] The assumption of asexual reproduction is made for expository convenience only; the formalism does not require it.

[11] For simplicity it helps to think of generations as non-overlapping, i.e. assume that the P-population goes out of existence as soon as the O-population comes into existence. But the formalism does not depend on this assumption.

Natural Selection in the Abstract

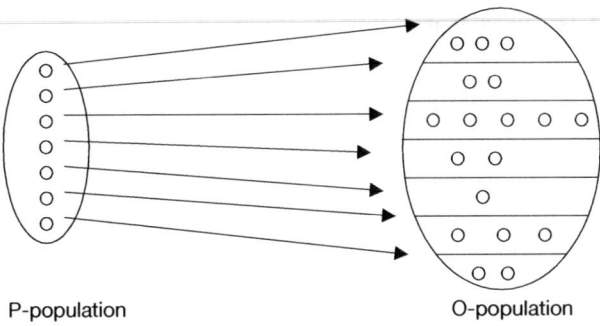

Figure 1.1. Relation between the P- and O-populations; n = 7

O-population, namely its average character value, the indices on the RHS of the formula refer to the P-population. In effect, we calculate average character value in the O-population by choosing an entity in the P-population, seeing what fraction of the O-population it is responsible for producing, and multiplying this fraction by the average character value of its offspring; we repeat this calculation for each member of the P-population, then take the summation. Figure 1.1 is a heuristic aid to seeing that this is a correct way of calculating $\bar{z}_o$.

The quantity we ultimately are interested in is $\Delta\bar{z}$, the change in average character value from one generation to another, where $\Delta\bar{z} = \bar{z}_o - \bar{z}$. Intuitively, it seems that $\Delta\bar{z}$ should depend somehow on the fitness differences in the P-population, and the fidelity with which the character z is transmitted. The Price equation captures this dependence precisely, by expressing $\Delta\bar{z}$ as the sum of two other quantities, as follows:

$$\bar{w}\Delta\bar{z} = \text{Cov}(w_i, z_i) + E(w_i \Delta z_i) \qquad (1.1)$$

Note that the quantity of interest, $\Delta\bar{z}$, appears on the LHS of equation (1.1) multiplied by average fitness $\bar{w}$, which is simply a normalizing constant. The first term on the RHS, $\text{Cov}(w_i, z_i)$, is the covariance between fitness w_i and character z_i. The second term on the RHS, $E(w_i \Delta z_i)$ is the average, or expected value, of the quantity $w_i \Delta z_i$, which is fitness x transmission bias. For ease of reading, we shall drop the indices wherever possible, so equation (1.1) can be rewritten:

$$\bar{w}\Delta\bar{z} = \text{Cov}(w, z) + E(w\Delta z) \qquad (1.1)$$

See Box 1.1 for the full derivation of equation (1.1).

Box 1.1. Derivation of the basic Price equation

$$\Delta \bar{z} = \bar{z}_o - \bar{z}$$

$$= \frac{1}{n} \sum_{1}^{n} \frac{w_i}{\bar{w}} z'_i - \frac{1}{n} \sum_{1}^{n} z_i$$

Multiplying both sides by $\bar{w}$ gives:

$$\bar{w} \Delta \bar{z} = \frac{1}{n} \sum_{1}^{n} w_i z'_i - \frac{1}{n} \sum_{1}^{n} \bar{w} z_i$$

Using the equality $z'_i = z_i + \Delta z_i$ to substitute for z'_i gives:

$$\bar{w} \Delta \bar{z} = \frac{1}{n} \sum_{1}^{n} w_i (z_i + \Delta z_i) - \frac{1}{n} \sum_{1}^{n} \bar{w} z_i$$

$$= \frac{1}{n} \sum_{1}^{n} w_i z_i - \frac{1}{n} \sum_{1}^{n} \bar{w} z_i + \frac{1}{n} \sum_{1}^{n} w_i \Delta z_i$$

$$= \frac{1}{n} \sum_{1}^{n} z_i (w_i - \bar{w}) + \frac{1}{n} \sum_{1}^{n} w_i \Delta z_i$$

Applying the standard statistical definitions of Covariance and Expectation gives:

$$\bar{w} \Delta \bar{z} = \text{Cov}(w_i, z_i) + E(w_i \Delta z_i)$$

which is the Price equation in its usual form.

A useful re-formulation of the Price equation results when we divide both sides by $\bar{w}$:

$$\Delta \bar{z} = \text{Cov}(\omega, z) + E_w(\Delta z) \qquad (1.2)$$

where Cov (ω, z) is the covariance between z_i and *relative* fitness ω_i rather than absolute fitness w_i; and $E_w(\Delta z)$ is the *fitness-weighted* average of the quantity Δz_i, rather than the simple average of the quantity $w_i \Delta z_i$, as in equation (1.1). We shall make use of both the absolute fitness and relative fitness formulations of Price's equation in what follows; obviously, it is easy to translate from one to the other.

1.3 INTERPRETATION OF PRICE'S EQUATION

What exactly does Price's equation mean? As we can see from equation (1.2), it expresses the total change in $\bar{z}$, between parent and offspring generations, as the sum of two other quantities. The first quantity, Cov (ω, z), measures the statistical association between the character z and fitness. If entities with a high character value tend to be fitter than average, then Cov (ω, z) will be positive; if such entities tend to be less fit than average, then Cov (ω, z) will be negative. If character value and fitness are completely unassociated, or if neither shows any variation at all, then Cov (ω, z) = 0. The covariance term is therefore a measure of the extent to which the character z is subject to natural selection; it is sometimes called the 'selection differential' on z.

The second quantity, $E_w(\Delta z)$, is a measure of the overall transmission bias in the population, weighted by fitness. To understand it, recall that *each* of the n entities in the P-population has a Δz_i term associated with it. $E_w(\Delta z)$ is the average of these n Δz_i terms, weighted by fitness. If each entity transmits its z-value perfectly, then $\Delta z_i = 0$ for each i, so $E_w(\Delta z) = 0$. However, if offspring deviate from their parents with respect to the character z, whether systematically or simply as a result of 'noise' during transmission, then $E_w(\Delta z)$ may be non-zero. Note that $E_w(\Delta z)$ is a fitness-weighted expectation: it takes into account not just how much the offspring of the ith entity deviate from it in character, but also how many offspring there are.

With these interpretations in mind, we see that the Price equation becomes highly intuitive. That $\Delta \bar{z}$ depends on Cov (ω, z) simply reflects the common-sense idea of natural selection—if taller organisms are fitter than shorter ones, that is, if height covaries positively with fitness, we expect average height in the population to increase over time. That $\Delta \bar{z}$ depends on $E_w(\Delta z)$ reflects the fact that transmission fidelity is important too—unless height is transmitted from parent to offspring with sufficient fidelity, then even if taller entities leave more offspring, average height will not necessarily increase. (Intuitively, this means that the magnitude of $E_w(\Delta z)$ should be related somehow to the heritability of z; see Section 1.5 below.) So Price's equation partitions the total change in $\bar{z}$ into two components, each of which has a natural biological interpretation.

If all fitness differences between entities stem from differences in survival, rather than fecundity, then the two components of Price's equation can be given a *temporal* interpretation. Viability selection leads the value of $\bar{z}$ to change *within* the P-population, between times t_1 and t_2; the magnitude of this change is given by Cov (ω, z). The surviving entities then reproduce, leading to a further change between times t_2 and t_3, of magnitude $E_w(\Delta z)$. So under pure viability selection, Cov (ω, z) equals the *within-generation* change in $\bar{z}$, while $E_w(\Delta z)$ equals the *subsequent* change that happens during the process of reproduction. But if there is a component of fecundity selection, this interpretation fails. Price's equation still holds true, of course, but the Cov and Exp components do not correspond to sequential periods of change.

A number of important points about Price's equation should be noted. First, the equation is simply a mathematical tautology whose truth follows from the definition of the terms. Nothing is assumed about the nature of the 'entities', their mode of reproduction, the mechanisms of inheritance, the genetic basis of the character, or anything else. It was Price's view that a properly general formalization of natural selection should abstract away from such contingent details (Price 1972, 1995; Frank 1995a). Rice (2004) observes that parental and offspring entities do not even have to be of the same type, so long as the character z is measurable on both, for example, parents could be groups and offspring organisms, or parents could be organisms and offspring gametes.[12]

Secondly, the character variable z can be defined however we please. To model the evolution of a 'discrete' rather than a 'continuous' character, for example, we simply need to define z appropriately. Suppose we are interested in the proportion of blue entities in our population, for some reason. We then define $z_i = 1$ if the ith entity is blue, $z_i = 0$ otherwise. Obviously, $\bar{z}$ then equals the proportion of blue entities in the P-population, and $\Delta \bar{z}$ the change in this proportion between the P- and O-populations. So Price's equation applies as usual. Similarly, z could be defined as the frequency of a particular allele at a given locus in an organism (= 1, $1/2$, or 0 for diploids); $\bar{z}$ would then equal the overall frequency of the allele in the population, and $\Delta \bar{z}$ the

---

[12] Indeed, the entities in the P- and O-populations do not need to be related as parents and offspring at all; as Price (1972) pointed out, his equation requires only an abstract correspondence between the two sets of entities.

change in frequency across one generation.[13] So Price's equation fits naturally with definitions of evolution such as change in gene frequency, or change in relative frequency of different types.

Where z denotes a continuous character, one might question whether all evolutionary change can be compressed down to $\Delta \bar{z}$, the change in the mean. For even if $\Delta \bar{z}$ equals zero, the character distribution may nonetheless have changed, for example, its variance may be different. Indeed in textbook cases of 'stabilizing selection', where extreme character values are selected against, the mean character remains the same from one generation to another but the variance is reduced; so only tracking the mean will create the illusion that no evolution has occurred. Though valid, this point compromises the generality of Price's equation less than it may seem. For if we wish, we can define z_i as the squared deviation in character of the ith entity from the population mean—which can be thought of as a relational property of the ith entity; $\bar{z}$ is then the variance of the character, and $\Delta \bar{z}$ the change in the variance, so Price's equation applies as usual. If z is suitably defined, the evolution of higher moments of the character distribution can be similarly captured.

Thirdly, note that Price's equation is statistical not causal. If Cov (ω, z) is non-zero, this means that differences in character value are correlated with differences in fitness, but the correlation need not reflect a direct causal link.[14] It is possible that z itself has no effect on fitness, for example, but is closely correlated with another character which does affect fitness. Where a non-zero value of Cov (ω, z) *is* due to a direct causal link between the character z and fitness, we shall say that z is *directly* selected; where Cov (ω, z) is non-zero for some other reason, then z is *indirectly* selected. The distinction between direct and indirect selection corresponds closely to Sober's (1984) distinction between 'selection of' and 'selection for'.

1.4 STATISTICAL VERSUS CAUSAL DECOMPOSITION

Despite its inherently statistical nature, the Price equation is often glossed in causal terms. Cov (ω, z) is often described as the component

---

[13] This allows standard population-genetic formulae for allele frequency change to be derived directly from Price's equation; see Michod (1999) p. 57 for an example.

[14] This point has been made repeatedly in relation to Price's equation and related formalisms, e.g. by Wade and Kalisz (1990), Heisler and Damuth (1987), Endler (1986), Lande and Arnold (1983), and Rice (2004).

of $\Delta \bar{z}$ 'due to' natural selection, while $E_w(\Delta z)$ is the component 'due to' transmission bias (e.g. Frank 1997). On this picture, the overall change in $\bar{z}$ is the net result of two separate causal factors, natural selection and transmission bias, whose relative magnitudes are given by the Price equation. The point noted above, that a non-zero value of Cov (ω, z) need not reflect a direct causal influence of z on fitness, complicates this picture somewhat, so let us shelve it for the moment: assume that z *does* causally influence fitness, and that no other character does. Is the causal gloss on Price's equation justifiable under these circumstances?

One reason for thinking not is that *both* terms of the Price equation, in its standard form above, contain the variable ω, denoting fitness. This suggests that the equation does not in fact resolve the total change into components due to selection and transmission bias respectively, for Cov (ω, z) and $E_w(\Delta z)$ are both affected by the fitness differences in the population. Intuitively, one would have thought that all the effects of fitness should be captured by the selection component. A simple re-formulation of equation (1.2) allows us to address this problem.

The re-formulation proceeds via an equation linking the weighted and unweighted expectations:

$$E_w(\Delta z) = E(\Delta z) + \text{Cov}(\omega, \Delta z)$$

This tells us that the fitness-weighted average of the Δz_i equals the simple average of the Δz_i plus the covariance between relative fitness ω_i and Δz_i. This last covariance measures the extent to which differences in fitness are associated with differences in transmission fidelity. Substituting this equation into equation (1.2) and rearranging gives:

$$\Delta \bar{z} = \text{Cov}(\omega, z') + E(\Delta z) \qquad (1.3)$$

This new form of Price's equation also expresses the total change as the sum of a covariance and an expectation, but with a difference. Cov (ω, z') is the covariance between an entity's relative fitness and the average character value of its *offspring*, rather than its own character value; $E(\Delta z)$ is the simple average of the parent–offspring character deviations, unweighted by fitness. Although Cov (ω, z') does not have as obvious a biological interpretation as Cov (ω, z), equation (1.3) is free from the shortcoming of (1.2) noted above. Only *one* of the terms on the RHS of (1.3) contains the term ω, which suggests that it does a better job than (1.2) in separating out the effects of natural selection.

Does equation (1.2) or (1.3) provide the correct decomposition of the total change? This question is worth thinking about both for its intrinsic interest and because the same *type* of question will arise again in later chapters. One might reply that the question is wrong-headed on the grounds that the two equations are different ways of describing the same thing, so the notion of 'correctness' does not apply. Frank (1998) defends a view of this sort. He notes that the total evolutionary change can always be partitioned into components in various ways; the choice between these alternative partitions, he says, 'is partly a matter of taste' (p. 12). Frank continues: 'The possibility of alternatives leads to fruitless debate. Some authors inevitably claim their partition as somehow true; other partitions are labelled false when their goal or method is misunderstood' (ibid. p. 12).[15]

To some extent I agree with Frank: it does not always make sense to try to choose between alternative statistical descriptions. However, I think there is a genuine distinction between statistical and causal decomposition, or partitioning. Equations (1.2) and (1.3) both provide correct *statistical* decompositions of $\Delta \bar{z}$, for both equations hold true by definition; but it still makes sense to ask which if either provides the correct *causal* decomposition. A brief digression is needed to explain this latter notion.

The issue of causal decomposition arises wherever a single effect is the result of more than one causal factor. In general, causal decomposition is only possible where the causal factors make 'separable' contributions to the overall effect (Northcott 2005). This will not always be the case. To borrow an example of Sober's, an individual's height is affected by both their genes and their nutritional intake, but we cannot ask how many centimetres are due to genes and how many to nutrition; this question makes no sense (Sober 1988). By contrast, in classical mechanics, if an object is acted on by two or more physical forces, then the overall effect, that is, the net acceleration, *can* be decomposed into components corresponding to each force, using standard vector analysis. So causal decomposition is sometimes but not always possible.

It is common in biology to regard the total evolutionary change in a population as the net result of a number of different causal

---

[15] Frank is not talking specifically about equations (1.2) and (1.3) in making these remarks.

factors, or 'forces', of which natural selection is one (Sober 1984).[16] Others include migration, drift, and transmission bias. This raises the question: is causal decomposition possible in this case? Can the total evolutionary change be divided into distinct components, each corresponding to a different causal factor? Biologists generally speak as if this *is* possible, for example, when they ask how strong selection in favour of a gene must be to prevent its elimination by drift, or how much migration is necessary to eliminate genetic variation between demes, or whether group selection in favour of a trait is strong enough to counter individual selection against it. Questions such as these presuppose that the relative magnitude of the different causal factors *can* be compared.

I think this presupposition is largely warranted. In general, it seems that causal decomposition should be possible if we can (i) identify the relevant causal factors, and (ii) for each factor, answer the question 'what would the effect have been if the factor had not operated?' (It is this last question that cannot be answered in Sober's height example above: it makes no sense to ask what height someone would have attained if they had had no genes, or received no nutrition at all.) By this criterion, total evolutionary change does turn out to be causally decomposable, at least in simple cases, though we shall encounter some complications in Chapter 3.

Now let us return to our question: which version of Price's equation, (1.2) or (1.3), provides the correct causal decomposition? (To facilitate comparison both equations are repeated below.) By hypothesis, the total evolutionary change in this case is affected by just two factors: natural selection and transmission bias. So we need to decide whether the change caused by selection equals $Cov(\omega, z)$ or $Cov(\omega, z')$. To do this, we have to ask what the total change *would* have been if there had been no selection on z. If the answer is $E_w(\Delta z)$, then the change due to selection equals $Cov(\omega, z)$, so (1.2) is the correct causal decomposition; if the answer is $E(\Delta z)$, then the change due to selection equals $Cov(\omega, z')$, so (1.3) is correct. For the actual change $\Delta \bar{z}$ must equal the change due by selection plus the change that would have occurred anyway, in

---

[16] Sober (1984) offers a philosophical defence of this way of thinking about evolution. His discussion is framed within a population genetics framework, where 'evolution' means the change in a population's allelic composition. The Price framework permits a generalization of this idea to all types of evolutionary change, not just changes in allelic composition.

the absence of selection.

	Change due to selection	Transmission bias	
$\Delta \bar{z} =$	$\text{Cov}(\omega, z)$	$+ \quad E_w(\Delta z)$	(1.2)
$\Delta \bar{z} =$	$\text{Cov}(\omega, z')$	$+ \quad E(\Delta z)$	(1.3)

There are various 'ways' in which there could have been no selection on z, that is, modifications of the actual world that eliminate selection on z. The simplest such modification, intuitively, is to equalize the fitnesses of all entities but leave everything else unchanged. The pattern of transmission is one of the things that stays unchanged, so Δz_i continues to describe the transmission bias of the ith entity. In this counterfactual scenario, the total change would be $E(\Delta z)$, not $E_w(\Delta z)$. For the entities are constrained to leave the same number of offspring as each other, and the offspring deviate from their parents in exactly the way they do in the actual world. So the total change would equal the simple average of the deviations $E(\Delta z)$, not the fitness-weighted average $E_w(\Delta z)$.[17] This suggests that in the actual world, the change due to selection equals Cov (ω, z'), not Cov (ω, z).

However, there are other ways to eliminate selection on z. For example, we could leave fitnesses unchanged but equalize character values, that is, make z_i the same for each i thus eliminating the phenotypic variance. If we presume that the pattern of transmission remains the same, the total change in such a world would be given by $E_w(\Delta z)$, not $E(\Delta z)$. For each entity would leave the same number of offspring as it does in the actual world, and those offspring would deviate from their parents just as in the actual world, hence the total change would be the fitness-weighted average of the deviations, or $E_w(\Delta z)$. If someone took 'the absence of selection on z' to refer to this counterfactual scenario, rather than the one above, they would equate the change due to selection with Cov (ω, z), not Cov (ω, z').

The choice between equations (1.2) and (1.3) now leads us into familiar philosophical territory. We want to know what $\Delta \bar{z}$ would have been if there had been no selection on z. But there are (at least) two 'ways' for there to be no selection on z, that is, modifications of the actual

---

[17] It is important to remember that the indices refer to the *actual* world. So although there is no variance in fitness in the imagined world, it does not follow that $E(\Delta z) = E_w(\Delta z)$, for the weights in the latter expectation are fitnesses of entities in the actual world, which do vary.

world that eliminate selection on z: (i) equalizing fitnesses and leaving everything else unchanged; (ii) equalizing character values and leaving everything else unchanged. In accordance with standard philosophical strictures, we need to decide whether (i) or (ii) constitutes a 'smaller' modification of actuality—for this determines how the counterfactual 'if there had been no selection on z' should be understood, which in turn determines how much of the actual change *is* attributable to selection on z.[18]

It seems clear that scenario (i) involves a smaller modification of actuality. For in both (i) and (ii), the transmission pattern is one of the things that remains unchanged. And while it is usually reasonable to assume that fitnesses can be equalized without affecting the pattern of transmission, it is *not* reasonable to assume this about character values. If all entities' character values are equalized, at whatever chosen level, then the pattern of transmission is most *un*likely to remain the same; this would occur only if character value were completely casually independent of deviation in character value between parent and offspring, which can hardly be the standard case.[19] So (ii) describes a scenario quite remote from the actual world.

This suggests that (1.3) rather than (1.2) provides the correct causal decomposition of $\Delta \bar{z}$. $E(\Delta z)$ equals the change due to transmission bias, which would have occurred anyway in the absence of selection; Cov (ω, z') equals the *additional* component of change due to natural selection. But equation (1.3) is only rarely found in the literature.[20] Indeed, it is quite common to find the components of equation (1.2) described as the 'change due to selection' and the 'change due to transmission' respectively.[21] But if I am right these labels should actually refer to the components of equation (1.3), not (1.2). What explains this discrepancy?

One possible explanation is that theorists reject the reasoning above, concerning the smallest modification of actuality that eliminates selection on z. But this seems unlikely: biologists rarely discuss such metaphysical matters. Another possibility is that theorists confuse causal

---

[18] Philosophical readers will recognize Lewis (1973) as the source of these ideas about causality and counterfactuals.

[19] The well-known phenomenon of 'regression to the mean', first noted by Francis Galton, highlights the implausibility of these two quantities being unrelated. Regression to the mean means, in effect, that the further a parent is from the mean of the parental character distribution, the greater the deviation in character between it and its offspring.

[20] Exceptions are Rice (2004), Frank (1997) equation 10, and Heywood (2005).

[21] For example, Frank (1997) says that 'one interpretation' takes these two terms to be the parts of total change 'caused by selection and transmission, respectively' (p. 1713).

with temporal decomposition. As we saw in Section 1.3, where selection is by differential survivorship alone, the components of equation (1.2) correspond to sequential periods of change—Cov (ω, z) equals the within-generation change in $\bar{z}$, while $E_w(\Delta z)$ equals the subsequent change that happens during reproduction. It may seem tempting to infer that Cov (ω, z) equals the change caused by selection, but this would be a fallacy. Even where selection is by differential survivorship alone, the *difference made by selection* cannot be equated with the within-generation change.[22] All the selection may take place in the parental generation, but its effect on the cross-generational change need not equal the within-generation change.

A final possibility is that biologists regard the difference between equations (1.2) and (1.3) as unimportant. But to accept this would require rejecting the distinction between statistical and causal decomposition altogether. As noted above, it is routine in biology to talk of evolutionary change as the result of different causal factors, of which natural selection is one; it is also routine to talk about the relative *strength* of selection versus other factors. If such talk is to be taken at face value, there must be an objective fact about how much of the change, in any particular case, is due to selection and how much due to other factors. So the possibility of causally decomposing the total change, and thus the distinction between statistical and causal decomposition, is not a philosophical extravagance; it is implicit in the way biologists actually speak.

Before leaving this topic, note that the question examined above, whether equation (1.2) or (1.3) provides the correct causal decomposition, does *not* violate the dictum that one cannot get causality out of correlation alone. No such feat was attempted: we explicitly stipulated that the character z *does* causally affect fitness, and that no other character does. Our question was: *given* these causal stipulations, how do we divide the total change into a component caused by selection and a component caused by other factors? Ascertaining whether the causal stipulations are true is another matter entirely.

1.4.1 Random Drift and Causal Decomposition

The role of stochastic factors was a major theme in twentieth-century evolutionary biology. I use the label 'random drift' as a shorthand for

---

[22] Except in the special case where Cov $(w, \Delta z) = 0$.

all such factors, though this somewhat extends its original, genetical meaning (Beatty 1992; Rice 2004). The Price formalism developed above does not take account of random drift. This is reflected in the fact that entity fitness w_i was defined as *actual* number of offspring produced—what is sometimes called 'realized fitness'. But if there is random variation in survivorship or fecundity, we need to distinguish realized from expected fitness; the latter is the number of offspring the entity would on average produce if it found itself in the same environment repeatedly. Differences in actual output *may* be due to differences in expected fitness; alternatively, they may be due to chance. The importance of this point has long been recognized; it is what motivates the 'propensity interpretation of fitness' (Mills and Beatty 1979; Sober 1984; Brandon and Carson 1996).

To incorporate drift into the abstract Price framework, we continue to define w_i as realized fitness, then decompose it into two parts. Suppose that in the environment in question, there is a range of numbers of offspring that the ith entity may leave, from zero to a maximum of m. Obviously, some of these possible levels of reproductive output will be more likely than others. So we can regard the entity's complete phenotype, plus the state of the environment, as inducing a probability distribution over these $m + 1$ possibilities, which we denote by P. Therefore, the expected fitness of the ith entity is:

$$w_i^* = \sum_{y=0}^{y=m} y \, P_y$$

where P_y is the probability of leaving y offspring in the environment in question. Obviously, on any one occasion realized fitness w_i may not equal expected fitness, because of chance factors. So we can write:

$$\underset{\text{fitness}}{\text{Realized}} \quad \underset{\text{fitness}}{\text{Expected}} \quad \text{Deviation}$$
$$w_i \quad = \quad w_i^* \quad + \quad \delta_i,$$

where δ_i is the deviation from expectation, which can be either positive or negative. The relative magnitudes of w_i^* and δ_i tell us how much of the ith entity's actual reproductive output is due to its expected fitness and how much due to chance.

We can then substitute the equality $w_i = w_i^* + \delta_i$ directly into Price's equation. Beginning with version (1.3) rewritten in terms of absolute fitness, and assuming for simplicity that there is no transmission bias, that is, $E(\Delta z) = 0$, this gives:

$$\overline{w}\Delta\overline{z} = \overset{\text{Change due to selection}}{\text{Cov}(w^*, z')} + \overset{\text{Change due to drift}}{\text{Cov}(\delta, z')} \quad (1.4)$$

Equation (1.4) partitions the total change into a component due to selection on z and a component due to random drift; it generalizes the causal decomposition provided by equation (1.3) to situations where drift operates. In principle, that is, if we could discover the probability distribution P, we could determine whether the overall change is the result of chance, natural selection, or a combination of the two.

This analysis sheds light on a recent debate in the philosophical literature. Matthen and Ariew (2002) argue that selection and drift should not be thought of as evolutionary 'forces' at all, on the grounds that there is no way of apportioning causal responsibility for a given evolutionary change between them; there is no 'common currency in which to compare the contributions of different evolutionary forces', they claim (p. 68). Equation (1.4) shows that this is not so. There is a common currency, namely units of mean character $\overline{z}$. It is straightforward (in principle) to compare the relative magnitudes of drift and selection in producing an evolutionary change; this is simply a matter of comparing Cov (w^*, z') with Cov (δ, z'). Price's equation thus provides a formal vindication of the 'force' metaphor that Matthen and Ariew oppose, by showing how to decompose the total change into components corresponding to each force, or cause. The issue of causal decomposition will resurface in Chapter 3, in relation to multiple levels of selection.

To analyse the evolutionary consequences of drift within the Price framework properly, it is necessary to consider the *expected* change in mean character due to selection, and thus the expected value of the covariance term Cov (w, z'). This analysis will not be pursued here (see Rice (2004) or Grafen (2000) for details.) For the purposes of addressing the levels-of-selection question, it will be simplest to ignore random drift altogether; so the distinction between realized and expected fitness will not be heeded in what follows.

1.5 PRICE'S EQUATION AND THE LEWONTIN CONDITIONS

We can use the version of Price's equation derived in Section 1.3—equation (1.3)—to forge a direct link with the Lewontin conditions for evolution by natural selection; this also sheds light on the relation between transmission bias and heritability.[23] A character's (narrow) heritability is traditionally defined as the linear regression of average offspring character on parental character, or mid-parental character if reproduction is sexual.[24] This is the definition we use here.

Figure 1.2 plots average offspring character z' against parental character z for a hypothetical population of entities. The plot is constructed as follows. For each parental entity, we ascertain its own character value z_i, and the average character value of its offspring z'_i; we then plot the point (z_i, z'_i) on the scatter diagram. So the diagram does not take account of fitness differences: each parental entity is represented by one point, irrespective of how many offspring it leaves. The heritability of the character z, which we shall denote by 'h', then equals the slope of the best-fitting regression line. By standard least-squares theory, it follows that $h = \text{Cov}(z', z)/\text{Var}(z)$, where the latter term is the variance of z.

With the regression line in place, each of the points in Figure 1.2 can be written as:

$$z'_i = a + h\, z_i + e_i$$

where a is the intercept, h the heritability, and e_i the unexplained residual, measuring deviation from the regression line.[25] Dropping the indices for convenience, we can now substitute this equation directly

---

[23] This section draws on ideas from both Queller (1992b) and Rice (2004), and on unpublished work by Ben Kerr.

[24] Narrow heritability can also be defined as additive genetic variance, i.e. the fraction of the phenotypic variance that is explained by the alleles acting additively, which under certain circumstances is equal to the offspring–midparent regression (cf. Roughgarden 1979). But offspring–parent regression is the more general concept, for shared genes are not the only cause of parent–offspring resemblance.

[25] Note that in writing this regression equation, we are *not* assuming that the 'true' functional relationship between z'_i and z_i is linear; it is possible to fit a best-fit regression line whatever the true functional relation looks like.

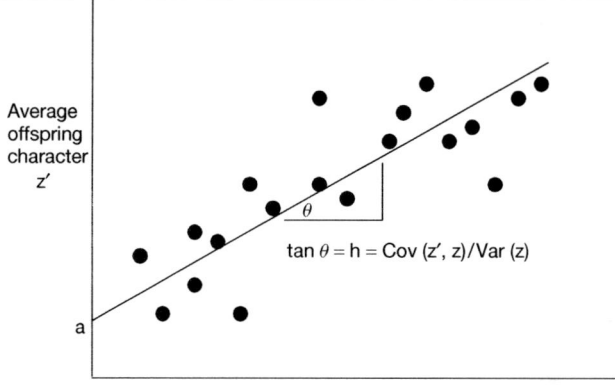

Figure 1.2. Offspring–parent regression for z

into the Price equation (1.3):

$$\Delta \bar{z} = \mathrm{Cov}\,(\omega, z') + \mathrm{E}(\Delta z) \qquad (1.3)$$
$$= \mathrm{Cov}\,(\omega, a) + \mathrm{Cov}\,(\omega, hz) + \mathrm{Cov}\,(\omega, e) + \mathrm{E}(\Delta z)$$
$$= h\,\mathrm{Cov}\,(\omega, z) + \mathrm{E}(\Delta z) + \mathrm{Cov}\,(\omega, e)$$

If we make the further assumption that $\mathrm{Cov}\,(\omega, e) = 0$, this simplifies to:

$$\underset{\text{Change due to selection}}{\Delta \bar{z} = h\,\mathrm{Cov}\,(\omega, z)} + \underset{\text{Transmission bias}}{\mathrm{E}(\Delta z)} \qquad (1.5)$$

The $\mathrm{Cov}\,(\omega, e) = 0$ assumption means that there is no correlation between fitness and scatter around the regression line, that is, fitness does not correlate with the residuals—the portion of offspring character that remains unexplained by regression on parental character. (See Queller (1992b) or Heywood (2005) for discussion.) This is a reasonable assumption since there is no reason why fitness *should* correlate with the residuals, which are usually thought of as 'random noise'. If the assumption is granted, equation (1.5) provides a useful decomposition of the total change. $\mathrm{E}(\Delta z)$, as we saw, equals the change that would have occurred anyway, without selection; hence $h\,\mathrm{Cov}\,(\omega, z)$ equals the change due to natural selection on z.

This is useful for two reasons. First, it provides a demonstration of a well-known result from quantitative genetics, often denoted $R = Sh$, where R is the 'response to selection', S is the selection differential, and h the heritability (Queller 1992b; Falconer 1989; Rice 2004). The selection differential is simply the within-generation change caused by viability selection, which as we have seen equals Cov (ω, z). Since h Cov (ω, z) is the total change due to selection, equation (1.5) proves this standard quantitative genetical result, presuming the 'response to selection' can be equated with the change caused by selection.

Secondly, equation (1.5) establishes a direct link with the Lewontin conditions for evolution by natural selection. Those conditions, to recall, were: (i) phenotypic variation; (ii) associated differences in fitness;[26] and (iii) heritability. Equation (1.5) tells us that the total change due to selection equals h Cov (ω, z). But by definition, Cov (ω, z) = Var (z) $b_{\omega z}$ where Var (z) is the variance of z, and $b_{\omega z}$ is the regression of ω on z. So if selection is to produce evolutionary change, then each of h, Var (z), and $b_{\omega z}$ must be non-zero, and these are precisely Lewontin's three conditions. Var (z) $\neq$ 0 means that z must vary—condition (i); $b_{\omega z} \neq 0$ means that the variation in z must be associated with fitness differences—condition (ii); and h $\neq$ 0 means that z must be heritable—condition (iii). So this version of Price's equation *appears* to provide a formal vindication of the Lewontin conditions, for it implies that Lewontin's conditions are individually necessary and jointly sufficient for there to be a component of change due to natural selection (see Box 1.2.)

Box 1.2. Lewontin's conditions and Price's equation

Total evolutionary change $\quad \Delta \bar{z} = \text{Cov}(\omega, z') + E(\Delta z)$

Change due to selection $\quad\quad\quad = \text{Cov}(\omega, z')$

$\quad\quad\quad\quad\quad\quad\quad\quad\quad\quad = h \text{ Cov}(\omega, z) \text{(assuming Cov}(\omega, e) = 0)$

$\quad\quad\quad\quad\quad\quad\quad\quad\quad\quad = h \text{ Var}(z) b_{\omega z}$

Lewontin's three conditions:
 (i) phenotypic variability, i.e. Var(z) $\neq$ 0
 (ii) associated differences in fitness, i.e. $b_{\omega z} \neq 0$
 (iii) heritability, i.e. h $\neq$ 0

---

[26] As noted in Section 1.1, this condition is often read causally, i.e. the phenotypic differences are meant to cause the fitness differences. Here I adopt a non-causal reading of Lewontin's second condition, i.e. phenotypic differences must correlate with fitness differences, in order to forge a link with the Price equation.

However, there is a slight complication. For equation (1.5) was derived with the help of the assumption that $\text{Cov}(\omega, e) = 0$, an assumption about which the Lewontin conditions say nothing. If this assumption is dropped, then it is possible for Lewontin's three conditions to be satisfied and yet for there to be *no* change due to selection, that is, if $h\,\text{Cov}(\omega, z) = -\text{Cov}(\omega, e)$. So strictly speaking, Lewontin's conditions are not sufficient for evolution by natural selection at all.

To understand what is going on here, note that what is *really* required for there to be evolution by natural selection is for $\text{Cov}(\omega, z')$ to be non-zero, that is, for an entity's fitness to correlate with the average character of its *offspring*. This is the fundamental condition, as is clear from equation (1.3). In effect, Lewontin's conditions try to replace this single condition with two others: a correlation between an entity's fitness and its *own* character, and a correlation between an entity's character and the average character of its offspring. (These two conditions are logically equivalent to Lewontin's three.) But this replacement is not always possible, for correlation, or covariance, is not in general a transitive relation. It is possible for x to covary with y, y to covary with z, but x not to covary with z. So even if character and fitness covary, that is, $\text{Cov}(\omega, z) \neq 0$, and even if the character is heritable, that is, $\text{Cov}(z, z') \neq 0$, it does *not* follow that $\text{Cov}(\omega, z') \neq 0$; and the latter is the fundamental condition that must be satisfied if selection is to lead to evolutionary change.

Is this a serious objection to the Lewontin conditions, or just a technicality? Certainly, it is most *unlikely* that Lewontin's conditions will be satisfied and yet selection not lead to evolutionary change. For as noted above, there is no particular reason why $\text{Cov}(\omega, e)$ should be non-zero, less still why it should exactly offset $h\,\text{Cov}(\omega, z)$. So for all practical purposes, Lewontin's conditions are correct. However, if logical precision is our goal, it is a mistake to treat Lewontin's conditions as sufficient for an evolutionary response to selection, as many authors do, for this flies in the face of the intransitivity of correlation. Overall, it seems fair to say that Price's equation *almost* vindicates the Lewontin conditions, by explaining why those conditions are an extremely good approximation, and identifying the very rare circumstances in which they fail. Where Lewontin's conditions are referred to in subsequent chapters, it will be assumed that these rare circumstances do not obtain.

Finally, let us consider the relation between transmission bias and heritability. If each entity 'breeds true', that is, transmits its character to its offspring with no deviation, then $\Delta z_i = 0$ for each i; this implies that $E(\Delta z) = 0$ and, so long as Var (z) $\neq 0$, that h = 1. So perfect transmission implies heritability of one, given non-zero variance in z. But the converse is *not* true. It is possible for heritability to be one and yet transmission not be perfect, in which case $E(\Delta z)$ may be non-zero. To illustrate, suppose that Var (z) $\neq 0$ and that Δz_i is a fixed, non-zero constant for each i, say 2. So each entity produces offspring which deviate from it by +2 units of character. It is easy to see that h = 1 in this scenario: parental deviation from the mean is a perfect predictor of offspring deviation from the mean. (In terms of Figure 1.2, the regression line fits all the points perfectly, has a slope of 1 and an intercept of 2, that is, $z' = 2 + 1(z) + 0$.) To put it differently, although entities do not transmit their characters perfectly, they do transmit their *character deviations from the mean* perfectly, hence heritability is 1, although $E(\Delta z) = 2$.

This example illustrates an important general moral. The inheritance mechanism at work in the population can affect the overall evolutionary change by affecting either the heritability h, or the transmission bias $E(\Delta z)$, or both. Figure 1.2 clarifies this point. The heritability of z equals the *slope* of the offspring–parent regression line, but as the example above shows, the transmission bias $E(\Delta z)$ depends also on the *intercept* of that line. Since the regression line must pass through the point $(\bar{z}, \bar{z}')$, and since by definition $E(\Delta z) = \bar{z}' - \bar{z}$, it is easy to see that:

$$E(\Delta z) = a + \bar{z}(h - 1)$$

which expresses the general relation between transmission bias and heritability. Thus the transmission bias *does* depend on the character's heritability, just as we would expect, but it also depends on the intercept of the regression line. Therefore, the heritability statistic does not embody all the information about the pattern of inheritance needed to determine the evolutionary change between generations.

One interesting consequence of this is that in theory, directional selection on a heritable character might lead to no evolutionary change overall, if the change due to selection is offset by an equal but opposite change due to transmission bias.[27] In terms of equation (1.5), the

[27] This point has been made by R. Brandon in an unpublished manuscript.

possibility is that h Cov $(\omega, z) = -E(\Delta z)$. Note that this is *not* an objection to the sufficiency of the Lewontin conditions; in such a circumstance selection *does* still produce an evolutionary response, it is just exactly offset by transmission bias. Selection still makes a difference: it causes the total change to be different from what it would have been had transmission bias been the only evolutionary force at work.

2
Selection at Multiple Levels: Concepts and Methods

INTRODUCTION

In Chapter 1, we noted that the Darwinian principles can potentially apply at more than one level of the biological hierarchy, perhaps simultaneously. This chapter looks in detail at the concepts needed to understand selection at multiple levels. Section 2.1 discusses the nature of hierarchical organization itself. Section 2.2 presents an abstract account of selection at two hierarchical levels, draws some preliminary distinctions, and contrasts two different types of multi-level selection. Section 2.3 shows how the Price formalism introduced in Chapter 1 can be extended to a hierarchical setting.

2.1 HIERARCHICAL ORGANIZATION

A typical representation of the biological hierarchy consists of a set of nested units, as in Figure 2.1, where each larger unit contains a number of smaller units.[1] This type of representation is useful, and it is easy to think of actual biological systems it could represent, for example, the three sizes of circle could depict multicelled organisms, eukaryotic cells, and intra-cellular organelles. But it is also misleading in its simplicity. Though the basic fact of hierarchical organization is not in dispute, there are competing ideas about how the biological hierarchy should be described, that is, which levels should be recognized and why. Moreover, it is not obvious which biological relation(s) are supposed to correspond to the abstract relation of containment depicted in Figure 2.1.

A detailed account of hierarchical organization has recently been developed by D. McShea (1996, 1998, 2001a, b). McShea is not

---

[1] I use the expressions 'unit' and 'entity' as stylistic variants.

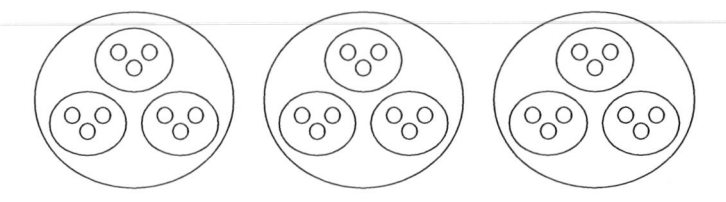

Figure 2.1. An abstract representation of hierarchical organization

directly concerned with levels of selection, but his ideas are nonetheless valuable in this context. His aim is to devise a scheme for understanding 'structural hierarchy', or part–whole structure, in nature. McShea argues that *interaction among the parts* is the key to the part–whole relation; it is this that distinguishes genuine wholes from arbitrary collections of lower-level units. So for example insect colonies count as genuine wholes, thanks to the behavioural interactions between the insects within a colony, even though the insects are not physically connected.

Treating interaction as the key to the part–whole relation sits well with many aspects of the levels-of-selection debate. A recurrent theme in the literature is that higher levels of selection come into play when lower-level units engage in fitness-affecting interactions with each another. In some cases, the part–whole structure that results from such interactions will coincide with 'physical' part–whole structure. For example, the interacting units may be physically connected to each other, like the cells in a multicelled organism, or be physically contained within a larger unit, like the organelles within a cell. But interaction is often regarded as fundamental. Thus for example, Sober and Wilson (1998) argue that if a group of organisms engage in no fitness-affecting interactions at all, hence have independent evolutionary fates, they do not constitute a 'real' biological group, whatever physical relations they bear to each other.

Importantly, not just any interaction-based part–whole relations generate a level of structural hierarchy in McShea's scheme. Multicelled organisms are made up of tissues and organs, which are themselves composed of cells, but McShea does *not* recognize tissues and organs as levels in between cells and organisms. This is because entities at all levels are required to be 'homologous with organisms in a free-living state, either extant or extinct' (2001b p. 408). This criterion is satisfied by cells and organelles, given accepted biological wisdom, but not by tissues or organs. Our kidneys, for example, are clearly not homologous to any free-living entities, extant or extinct. Most characterizations of the biological hierarchy implicitly rely on this free-living existence requirement; without it, there would be virtually no limit on the hierarchical depth we could discern in biological systems.

McShea's scheme is well-suited for thinking about 'evolutionary transitions' in the sense of Maynard Smith and Szathmáry (1995) and Michod (1999)—an important aspect of the levels-of-selection discussion.[2] Evolutionary transitions occur when a number of lower-level entities, originally capable of surviving and reproducing alone, become aggregated into a larger unit. Clearly, the hierarchical structure generated by such transitions automatically satisfies the free-living existence requirement. Furthermore, entities at all levels in McShea's hierarchy are in principle capable of reproduction, so can evolve by natural selection. Of course, some lower-level entities lose this capacity as they become integrated into a larger unit, for example, sterile workers in an insect colony. But the requirement of homology to a free-living entity means that entities at all levels are the *sorts* of thing that can reproduce, in the way that bodily organs, for example, are not. This in turn provides a partial justification for the strategy adopted in the previous chapter, of treating reproduction as a primitive concept.

One potential objection to McShea's scheme is that his account of the part–whole relation is too liberal. Many theorists would agree that interaction among lower-level units is *necessary* for them to constitute a whole, but not that it is sufficient. For example, Michod and Nedelcu (2003) say that 'the transition to a new higher level is driven by the interaction among lower-level units'; but they hold that a group of lower-level units only constitutes an 'evolutionary individual' once conflict-reducing mechanisms have evolved to prevent within-group competition undermining the integrity of the whole (p. 64). Similarly, Szathmáry and Wolpert (2003) argue that 'overall co-ordination of function' is required for a colony of cells to qualify as a multicelled individual, noting that this rules out most bacterial colonies (p. 273). So on this view, interaction among parts is not sufficient for the existence of a whole; functional organization is also required.

---

[2] Though McShea himself sees significant differences between his structural hierarchy and the type of hierarchical organization that Maynard Smith and Szathmáry (1995) are concerned with. I follow Sterelny (1999) in regarding these differences as less significant than McShea thinks.

A similar issue arises in the group selection literature. Uyenoyama and Feldman (1980) and D. S. Wilson (1975) both defined groups in terms of fitness-affecting interactions among organisms—what Wilson called 'trait groups'. Sober and Wilson (1998) offer a sustained defence of the trait-group concept, embracing the consequence that the lifespan of a group may be shorter than that of its members. Any two organisms engaged in a fitness-affecting interaction, cooperative or competitive, constitute an evolutionary group, they argue, irrespective of the duration of the interaction. But critics regard the trait-group concept as excessively liberal (e.g. Nunney 1998; Maynard Smith 1998; Sterelny 1996b). These critics typically require some degree of group-level functional organization, for example, division of labour or differentiation into castes, for the existence of groups as meaningful evolutionary units.

How should this issue be resolved? One point in favour of the McShea/Sober and Wilson position is that functional organization is presumably an adaptation of groups, hence something that might *result* from a process of group selection. Since group selection can only occur if there actually are groups to select, the existence of groups should not be made contingent on properties that can only evolve by group selection, or a chicken-and-egg problem arises (cf. Williams 1992). This suggests that the criteria by which hierarchical levels are defined should be relatively liberal, that is, interaction *should* be treated as sufficient for part–whole structure. Features such as division-of-labour and differentiation among parts can then evolve by selection on wholes that already exist, avoiding the chicken-and-egg problem.

However, the McShea/Sober and Wilson position faces one serious problem. Typically, the biological hierarchy is represented as strictly nested: an entity at level X belongs to exactly one entity at level $X + 1$, as in Figure 2.1. But if we define higher-level entities on the basis of interactions among lower-level entities, there is no guarantee that the resulting hierarchy *will* be strictly nested.[3] To see why, imagine a large population of lower-level entities that engage in fitness-affecting interactions. It is possible that entity x interacts with y, y interacts with z, but x does not interact with z. This means that we cannot partition the population into non-overlapping groups using just the criterion of interaction. We can still discern groups in the population, but they will overlap, as shown in Figure 2.2. If we treat entity x as focal, we find an

---

[3] See Godfrey-Smith (forthcoming) for an extended discussion of this problem.

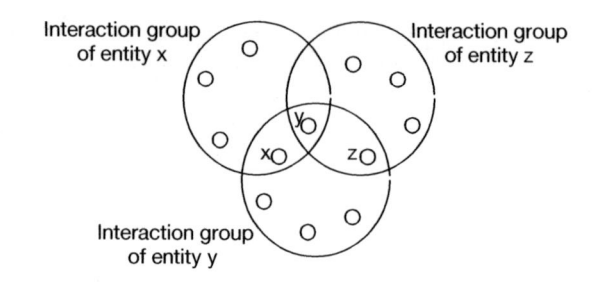

Figure 2.2. Overlapping hierarchical structure

interaction group that includes y but not z; treating entity y as focal, we find an interaction group that includes both x and z. So the principle of strict nesting is violated.[4]

There are two possible responses to this problem, neither of which is wholly satisfactory. The first is to impose strict nesting as a *sui generis* requirement on structural hierarchy, rather than hoping it will emerge from the pattern of interaction. In effect, this is to regard interactions between parts as insufficient for the existence of a whole; the latter requires, in addition, that all interactions be transitive. The second is simply to drop the requirement of strict nesting, that is, to allow that the principles of hierarchical organization can generate either a nested or a non-nested set of units.[5] On this approach, interactions among parts can be retained as sufficient for the existence of a whole; it just becomes contingent whether the resulting hierarchy is nested or not.

I think the second solution is on balance preferable. For as noted above, on one standard conception of multi-level selection, higher levels of selection arise via fitness-affecting interactions between lower-level units. Though models of this process often assume a nested hierarchy, the underlying causal mechanism does not require nesting; it could work equally well with overlapping groups of lower-level units. So relative to this conception, it would a mistake to insist that the hierarchy be strictly

---

[4] Formally, this is because the relation 'x interacts with y' is non-transitive, while strict nesting requires that the relation 'x is in a group with y' be transitive. So the two relations can only be co-extensive if the hierarchy is allowed to violate the assumption of strict nesting.

[5] This issue should not be dismissed as academic on the grounds that the actual biological hierarchy *is* strictly nested. Rigorous application of Wilson's trait-group concept would almost certainly imply the existence of overlapping groups in nature. Similarly, given standard ways of individuating ecosystems, many populations may turn out to belong to more than one ecosystem (Sterelny 2006). It is only at the lowest hierarchical levels that nesting appears perfectly strict.

nested. This consideration is not decisive, since the conception of multi-level selection in question is controversial, as we shall see. But it does show that the problem of characterizing the biological hierarchy cannot be divorced from the problem of characterizing multi-level selection; they are interdependent.

The overlapping problem raises the worry that interactions among parts may not suffice for the existence of wholes. A different worry is that interactions may not be necessary. There are biological units that contain parts, hence intuitively count as hierarchically structured, but where the parts engage in little if any interaction with each other. *Genets* are a good example. A genet refers to the totality of plant tissue that comes from a fertilized zygote. In plants with highly clonal habits, a genet may be composed of a large number of ramets, or physiologically discrete plants, produced vegetatively. In some clonal species the ramets in a genet disperse widely, to lead wholly separate lives.[6] And even if all the ramets in a genet do interact, it is not *because* of this that they belong to the same genet. So the ramet–genet case is a part–whole relation not definable in terms of interaction.

Of course, one might deny that the ramet–genet relation *is* part–whole, in the salient sense. But a more plausible response would be to follow Eldredge (1985) in distinguishing the ecological from the genealogical hierarchy. In the former, smaller units form parts of larger units in virtue of their ecological interactions; in the latter, it is genealogical relations that bind the smaller into the larger units. The two hierarchies are parallel, according to Eldredge, though multicelled organisms belong in both.[7] McShea's ideas pertain exclusively to the ecological hierarchy, as do those of theorists interested in evolutionary transitions.[8] But the ramet–genet relation pertains to the genealogical hierarchy: ramets belong to the same genet in virtue of deriving from the same zygote. Similarly, the relation between an organism and the species to which it belongs is genealogical, not ecological. So genets and species are levels in the genealogical hierarchy.[9]

---

[6] Though in other species the ramets remain tightly connected, physically and physiologically, after propagation. Most clonal species lie somewhere between these two extremes.
[7] This is because the cells within a multicelled organism both interact with each other, and are united by their descent from a single zygote.
[8] As McShea himself notes (2001a, p. 505).
[9] Note that if the genealogical hierarchy is constructed using a strict criterion of monophyly, then the resulting hierarchy must be strictly nested. For monophyletic

Should a characterization of hierarchy adequate for addressing the levels-of-selection question admit part–whole relations that are genealogically defined? The answer is surely yes. For the concept of reproduction, hence fitness, applies to at least some genealogically defined entities. Species beget daughter species when they speciate, and genets beget daughter genets when their constituent ramets reproduce sexually rather than clonally, so both can evolve by natural selection. Though species selection is a controversial idea, the same is not true of genet-level selection. Indeed the traditional way of applying Darwinian theory to clonal plants treats genets as the bearers of fitness.[10] Vegetative propagation, on this view, constitutes growth not reproduction (Janzen 1977). But this idea has not gone unchallenged. Some theorists favour ramet-based definitions of fitness; while others favour a *multi-level* approach, in which fitness is attributed to both ramets and genets (e.g. Tuomi and Vuorisalo (1989), Pedersen and Tuomi (1995), Monro and Poore (2004)). Clearly, this application of multi-level selection theory requires a conception of hierarchy that permits genealogically defined part–whole relations.

To sum up, hierarchical organization can arise in various ways. Interaction, physical connectedness, and common descent are biological relations that can bind smaller units into larger ones, generating part–whole structure. In some cases the three relations coincide, but often they do not. McShea's scheme treats interaction as all-important; it fits many aspects of the levels-of-selection debate, but not all. Entities belonging to Eldredge's genealogical hierarchy can also evolve by natural selection.

2.2 SELECTION AT MULTIPLE LEVELS: KEY CONCEPTS

More could be said about hierarchical organization itself, but the foregoing is enough to allow us to consider selection in a hierarchical context. Our aim is to extend the abstract characterization of natural selection developed in the previous chapter to a hierarchical or multi-level setting. In principle this should be straightforward, and in some

---

groups, of necessity, are strictly nested within each other. So if the genealogical hierarchy is a hierarchy of monophyletic inclusion then it is necessarily non-overlapping.

[10] Pan and Price (2002), discussing Darwinian approaches to clonal evolution, say that genets 'have traditionally been viewed as the appropriate units of selection, and, thus, for measuring fitness' (p. 584).

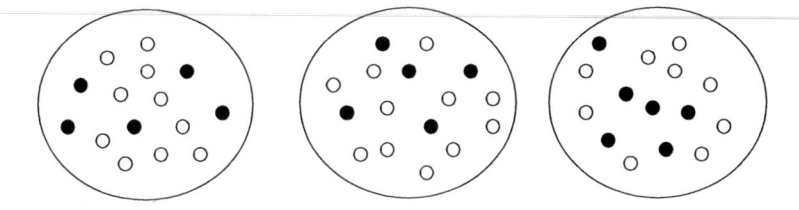

Figure 2.3. Particles of two types nested within collectives

ways it is. However, multi-level selection raises a number of conceptual and theoretical ambiguities, some of which are quite subtle.

For simplicity, we consider a two-level scenario with strict nesting, as depicted in Figure 2.3; it may be useful to keep this diagram in mind. We shall refer to the lower-level units as 'particles' and the higher-level units in which they are nested as 'collectives'.[11] These designations are strictly relative: the hierarchy may extend downward or upward. Relative to still lower levels, our particles would count as collectives; relative to higher levels, our collectives would count as particles. Focusing on just two levels simplifies the analysis; the extension to further levels raises no new issues of principle. The assumption of strict nesting is relaxed later.

Recall the basic requirements for evolution by natural selection: character variation, associated differences in fitness, and heritability. We saw in the previous chapter how the cross-generational change in mean character, in a one-level scenario, depends on these three factors. Intuitively, if selection is to operate at multiple levels, and lead to evolution, then entities at each level must satisfy the three requirements. Thus in a two-level scenario, the particles must vary with respect to a heritable character and differ in fitness as a result; and similarly for the collectives. If this is right, then the essence of multi-level selection is the *simultaneous* existence of character differences, associated differences in fitness, and heritability at more than one hierarchical level.

This raises an overarching question: what is the relation between the characters, fitnesses, and heritabilities at each level? For example, how does the fitness of a collective relate to the fitnesses of the particles within it? Does variance at the particle level necessarily give rise to variance at the collective level? Does the heritability of a collective character depend somehow on the heritability of particle characters? The literature on

---

[11] 'Particle' comes from Hamilton (1975), 'collective' from Kerr and Godfrey-Smith (2002).

multi-level selection has rarely tackled these questions explicitly, but they are crucial.

2.2.1 Particle Characters and Collective Characters

Consider characters first. Suppose that the particles vary with respect to a measurable character, denoted z. What does this imply about the characters of the collectives in which the particles reside? Intuitively, we can distinguish between collective characters that derive directly from underlying particle characters, and those that do not. An example of the former is 'average z-value of the particles in a collective', or Z. Clearly, this collective character is logically determined by the particle characters—fix the z-value of each particle and the Z-value of each collective is automatically fixed too. Such characters often play a role in evolutionary models. For example, in many group selection models the salient group character is 'average gene frequency in the group', which derives directly from the gene frequencies of the organisms in the group. Collective characters of this sort are sometimes called 'aggregate', because they are produced by aggregating the characters of the particles in the collective (Vrba 1989; Lloyd 1988; Grantham 1995).

Other collective characters are not such simple functions of particle-level characters, though they usually depend on them indirectly.[12] For example, the degree of morphological differentiation between castes, or the number of cell divisions before germ-line sequestration, are characters of insect colonies and multicelled organisms, respectively, that bear a much less direct relation to the characters of the particles—insects and cells—that compose the collectives. Such characters are sometimes called 'emergent'. Emergent characters can be relevant to selection too; indeed, some theorists argue that selection at any level *requires* emergent characters at that level, though I will argue against this.

The aggregate/emergent distinction, though intuitive, is hard to make precise and has been characterized in non-equivalent ways in the literature.[13] I suspect that this is because aggregate and emergent are

[12] Usually but not necessarily. Since a collective may be physically composed of things in addition to particles, some of its properties may fail to supervene on the properties of its constituent particles. Note that this does *not* imply a violation of the metaphysical thesis known as 'mereological supervenience', which says that the properties of wholes are fully determined by the properties of their parts. The point is that a collective's 'parts' may include things other than the lower-level particles.

[13] For competing ideas on how to characterize emergent properties, see Salt (1979), Vrba (1989), Eldredge (1989), and Grantham (1995). Williams (1992) is sceptical about

really opposite ends of a continuum, along which the determination of collective by particle characters ranges from the simple to the complex, rather than dichotomous alternatives. For the moment, the point to note is just that particle and collective characters can be related in various ways.

It is sometimes useful to regard collective characters as 'relational' or 'contextual' characters of the particles in the collectives (Damuth and Heisler 1988). For example, if a particular ant colony has a female-biased sex ratio, then the ants within the colony have the property of *belonging* to a colony with a female-biased sex ratio. This may sound trivial, but it plays an important role in certain conceptualizations of multi-level selection, as we shall see. Its importance lies in the fact that a particle's fitness may be affected by characters of the collective to which it belongs.

2.2.2 Life Cycles

To understand simultaneous fitness attributions to particles and collectives, we need to consider their life cycles, and mode of reproduction. In a single-level scenario one can in principle study selection without consideration of entities' life cycles, by restricting attention to viability selection and distinguishing sharply between selection itself and the evolutionary response to selection. Modulo this restriction, it makes no difference what sort of life cycle the entities have, nor how they reproduce—the selection stage is the same whatever. But in a multi-level scenario, reproduction cannot be hived off so neatly from selection, even if we restrict attention to viability selection.[14] For the life cycles of entities and particles may not be synchronized.

Synchronization of life cycles means that particle and collective generations go hand-in-hand: reproduction of particles is accompanied by reproduction of collectives and vice versa. Chromosomes and cells provide an example. Chromosomal replication usually occurs once during each cell cycle, just before cell division, so there are as many generations of cells as there are of chromosomes.[15] The same is true of

---

the aggregate/emergent distinction; Heisler and Damuth (1988) are sceptical about its significance. Philosophers of psychology have also discussed the notion of an emergent property; see Beckerman, Flohr, and Kim (eds.) (1992) and Clayton and Davies (eds.) (2006).

[14] Indeed Michod (1999) argues that the very distinction between selection and the response to selection breaks down when we consider multi-level scenarios.

[15] However in 'coenocytic' organisms such as fungi, cellular division and nuclear division are decoupled, with the result that the fungal mycelium may contain a large number of nuclei sharing a common cytoplasm.

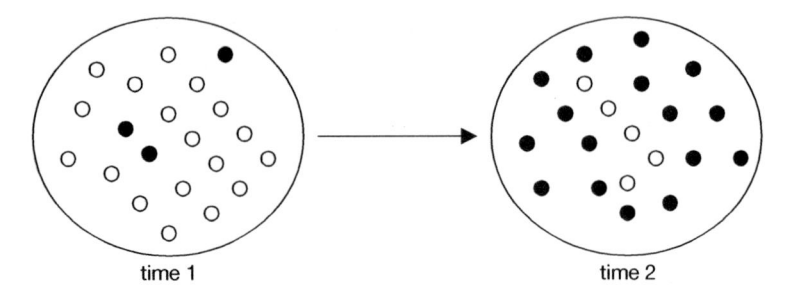

Figure 2.4. Particle-level selection within the lifespan of a collective

the individuals and groups in D. S. Wilson's (1975) trait-group model, where the groups assemble and break up every organismic generation. Contrast this with the case of mitochondria and cells, or cells and multicelled organisms. Mitochondria can divide many times during a single cellular generation, just as cells divide many times within a single organismic generation. So in these cases, the life cycles of particles and collectives are not synchronized.

Where generations are non-synchronized, different particle-types may change in frequency within the lifespan of a single collective, owing to differences in their rate of survival or reproduction; see Figure 2.4. This is one possible meaning of 'particle-level selection'. An example is the 'somatic' selection in the immune system, where T cells with a given antibody specificity become more abundant within the lifespan of their host organism. Similarly, if the particles are organisms and the collectives are demes, then differential reproduction of variant organism-types within a deme's lifespan constitutes selection of the same sort. Conceptually, the relation between collective and particle is straightforward here: the collective is simply the 'arena' in which the particle-level selection is played out.

Somatic selection is sometimes dismissed as evolutionarily inconsequential on the grounds that the selected variants are not transmitted to the next (organismic) generation. But as Buss (1987) notes, this is true only for taxa in which germ-line cells are sequestered in early ontogeny. More generally, whether selection between particles within a collective's lifespan has long-term consequences depends on how the collectives reproduce. There are a variety of modes of collective reproduction. Collectives may reproduce asexually, for example, by fissioning, or by emitting small propagules which then develop. Sexual modes include

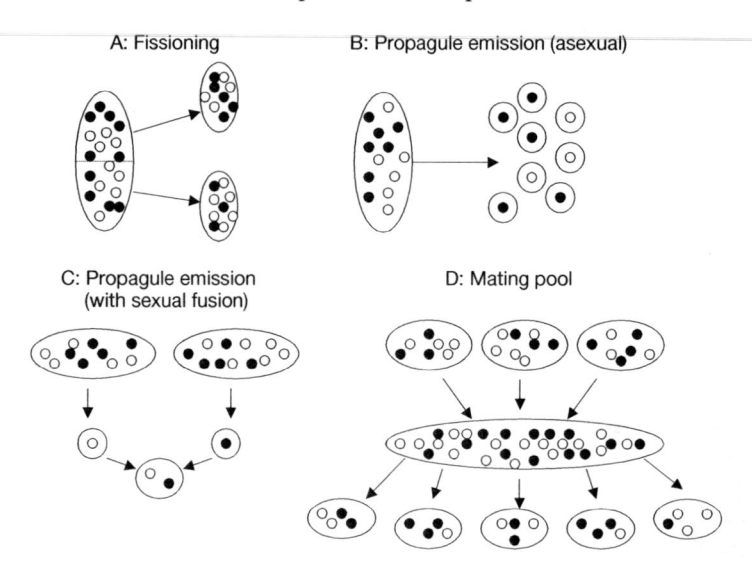

Figure 2.5. Modes of collective reproduction

emitting propagules which fuse with other propagules before development, or contributing particles into a 'mating pool' which then aggregate to form the next generation of collectives (cf. Wade 1978). These modes of reproduction, all of which are found in nature, are illustrated in Figure 2.5.

Where collectives are formed by aggregation, the notion of collective reproduction in the sense of collectives making more *collectives* can become strained. Consider for example the slime mould *Dictyostelium*, which is formed by aggregation of a large number of single-celled amoebae (Figure 2.6). Usually these amoebae have a free-living existence and reproduce asexually. But when food is scarce, between 10^4 and 10^6 of them come together to form a mobile slug. Once the slug has moved some distance through the soil, the amoebae differentiate into a sterile stalk and a fruiting body that produces spores for dispersal; the life cycle then returns to the single-celled state. Slime moulds are undoubtedly 'real' entities, with many of the characteristics of multicelled organisms; but it would be virtually impossible to identify parent–offspring lineages among the slime moulds themselves. For the amoebae that aggregate to form a slime mould are large in number and may be of quite different ancestries; so any given slime mould may have an indefinitely large

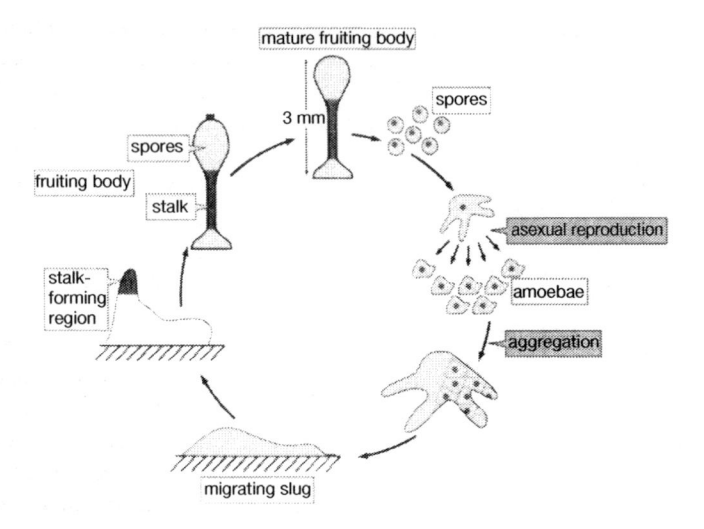

Figure 2.6. Life cycle of the cellular slime mould [based on Figure 12.4 of Maynard Smith and Szathmáry 1995]

number of 'parent' slime moulds, of very different ages. In this case, collective reproduction is more easily understood in terms of production of offspring *particles*, not collectives.

Where generations are *not* synchronized, it is possible to 'restore' synchrony by computing the fitness of a particle across a timescale longer than one generation (cf. Rice 2004 ch. 10). For example, if five rounds of particle reproduction correspond to one round of collective reproduction, we can define a particle's fitness as the number of offspring it leaves after five particle generations, rather than one. In this way the intrinsic difference in turnover rate between particles and collectives is exactly compensated for, rendering the effects of selection at the two levels commensurable. This is perfectly legitimate; defining fitness in terms of immediate offspring is not mandatory. Some evolutionary phenomena require us to consider offspring counts over more than one generation, as Fisher (1930) noted in his treatment of sex-ratio evolution.

As an example of restoring synchrony in this way, consider the well-known idea that meiotic drive constitutes a form of lower-level or 'genic' selection that can operate alongside genotypic selection in a one locus population-genetic model (Wilson 1990; Kerr and Godfrey-Smith 2002; Okasha 2004a). In this scenario, the particles are alleles and the

Selection at Multiple Levels: Concepts and Methods 53

collectives are diploid organisms ('genotypes')—which can be thought of as groups of two alleles. So there are two levels of selection—between alleles within a genotype, and between genotypes. Since many rounds of DNA replication occur within an organism's lifespan, generations are obviously not synchronized. However, particle fitness is computed across the generation-time of the collective: an allele's fitness is defined as the number of descendants it leaves in the next *organismic* generation (equivalently, the number of copies of the allele found in the organism's successful gametes.) In effect, this way of defining allelic fitness converts the multi-level scenario into one with synchronized generations.[16]

Confusion can arise from switching between long- and short-term definitions of particle fitness. For example, in *The Evolution of Individuality*, Buss (1987) describes how variant cell lineages within an organism may compete for access to the germ line. Cellular variants which stop making somatic tissue in order to gain access to reproductive tissue 'fail to behave altruistically', he argues: they selfishly pursue their own interests at the expense of the whole organism (p. 103). But when he discusses germ-line sequestration later in the book, Buss says exactly the opposite. Sequestered germ cells are mitotically inactive, he notes, hence 'losers in cell lineage competition. The selfish cells here are the *somatic* cells, which abandoned a function of significance to the individual, in return for further replication' (p. 180, my emphasis). This contradiction is the result of Buss changing the timescale over which the relative fitness of the different cell-types is being considered. Somatic cells leave more descendants than germ cells *within* the lifespan of the organism, but leave no descendants across organismic generations. So deciding which cell-type is fitter, or has a higher replication rate, depends on which timeframe we consider.

2.2.3 Particle Fitness and Collective Fitness

To simplify the treatment of fitness at two levels, we ignore the distinction between realized and expected fitness. The notion of particle fitness is then straightforward. A particle's fitness is the number of offspring particles it leaves, over some appropriate number of particle

---

[16] Note that 'allelic fitness' as the term is used here means the fitness of an allele within a diploid genotype. This is *not* the same as the 'marginal allelic fitness' of standard population genetics textbooks, which is the fitness of an allele-type averaged across all genotypes. See Chapter 5, Section 5.3 and Kerr and Godfrey-Smith (2002) for further discussion of this point.

54   *Evolution and the Levels of Selection*

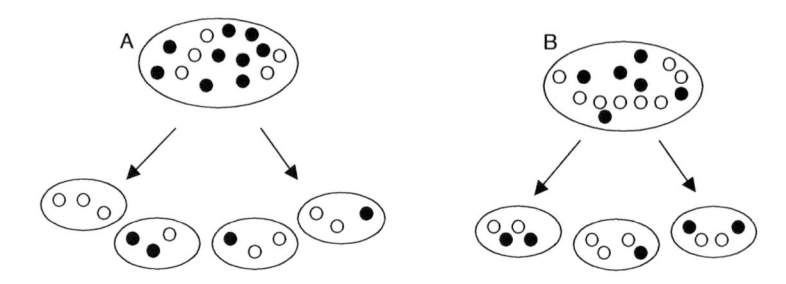

Figure 2.7. Non-equivalence of collective fitness$_1$ and fitness$_2$

generations, perhaps one. Collective fitness, however, is less straightforward. As has often been noted, there are two ways that collective fitness could be defined (cf. Arnold and Fristrup 1982; Damuth and Heisler 1988; Sober 1984; Rice 2004). First, a collective's fitness could be defined as the average or total fitness of its constituent particles; so the fittest collective is the one that contributes most offspring *particles* to future generations of particles. Secondly, a collective's fitness could be defined as the number of offspring *collectives* it leaves; so the fittest collective is the one that contributes the most offspring collectives to future generations of collectives. I call these concepts 'collective fitness$_1$' and 'collective fitness$_2$' respectively.

The difference between these two concepts is illustrated in Figure 2.7. Two collectives, containing equal numbers of particles, reproduce asexually. Collective A produces four daughter collectives while B produces three. However, the total number of offspring particles they produce is identical: twelve. So A has greater fitness$_2$ than B—it contributes more collectives to the next generation. But A and B have the same fitness$_1$—they contribute equal numbers of particles to the next generation. Note that in computing fitness$_2$, we count offspring collectives *without* weighting by size; if we weighted by size, we would in effect be counting the particles themselves, hence computing fitness$_1$. The difference between the two concepts of fitness is clearest when the collectives reproduce asexually, but also holds for sexual reproduction.

In some cases the two types of collective fitness will go hand in hand. One way this can occur is as follows. In Figure 2.7, A's offspring collectives start life containing three particles, versus four for B's offspring. So it is possible that the latter will attain a larger adult size, and thus leave more offspring collectives themselves. If so, A and B will leave equal

Selection at Multiple Levels: Concepts and Methods   55

numbers of *grandoffspring* collectives. So if collective fitness$_2$ is judged over two generations, rather than one, A and B come out equally fit. Put differently, A's fitness advantage after one round of reproduction is exactly offset by the reduced fitness of its offspring, meaning there is no intrinsic advantage to producing lots of small offspring rather than a few larger ones, *even though* we are counting unweighted. A collective's fitness$_2$ thus becomes directly proportional to its fitness$_1$. This scenario may seem unlikely, but it plays a role in Michod's theory of the evolution of multicellularity, discussed in Chapter 8 (Michod 1999; Michod and Roze 1999).

The fact that collective fitness$_1$ will sometimes be proportional to fitness$_2$ does not undermine the importance of the distinction, for two reasons. First, there is no necessity that this will be so—it depends on contingent biological facts. Secondly, the *absolute* fitness$_1$ of a collective is always different from its absolute fitness$_2$, for these quantities are measured in different units: number of offspring particles and number of offspring collectives, respectively. So while the *relative* fitness of a collective in the two senses may be identical—relative fitness being a dimensionless quantity—the conceptual difference between the two types of collective fitness remains.

Most formal models of group selection have defined 'group fitness' the first way, as average fitness of the individuals in the group. This is because the aim of such models is to understand the evolution of an *individual* phenotype, often altruism, in a population that is subdivided into groups (Damuth and Heisler 1988; Okasha 2001; Krause and Ruxton 2002). By contrast, in the macroevolutionary literature on species selection, 'species fitness' has been defined the second way, as expected number of offspring species. This is because species selection is meant to explain the changing frequency of different types of species, not organisms. So the fittest species are not the ones whose constituent organisms are especially well-adapted, but rather the ones with the greatest probability of surviving and speciating (i.e. reproducing). It follows that species selection is *not* simply a higher-level analogue of traditional group selection, as Arnold and Fristrup (1982) note, for it is of a different logical type.

The distinction between fitness$_1$ and fitness$_2$ is really a special case of the distinction between aggregate and emergent characters. (An entity's fitness can always be regarded as one of its phenotypic characters.) Clearly, collective fitness$_1$ is aggregate—it arises directly from the characters of the particles within the collective. Fix the fitness of every particle in the population, and the fitness$_1$ of every collective is fixed too. But collective fitness$_2$ is emergent: it bears no logical or definitional relation to the characters of the particles within the collective. Even if we knew the fitness and complete phenotypic description of every organism in a species, for example, we could not necessarily make any prediction about the species' chance of surviving and speciating.

2.2.4 The Two Types of Multi-Level Selection

I turn now to an important ambiguity in the concept of multi-level selection, discussed by authors including Arnold and Fristrup (1982), Sober (1984), Mayo and Gilinsky (1987), Damuth and Heisler (1988), Okasha (2001) and others. The ambiguity arises because there are two things that multi-level selection can mean, that is, two ways that the basic Darwinian principles can be extended to a hierarchical setting. The ambiguity has usually been discussed in relation to individual and group selection, but it generalizes to any multi-level scenario, for it stems from the distinction between collective fitness$_1$ and fitness$_2$.

Consider again the scenario depicted in Figure 2.3, where a number of particles are nested within each collective. The key issue is whether the particles or the collectives (or both) constitute the 'focal' level.[17] Are we interested in the frequency of different particle-types in the overall population of particles, which so happens to be subdivided into collectives? If so, then the particles are the focal units; the collectives are in effect part of the environment. Alternatively, we may be interested in the collectives as evolving units in their own right, not just as part of the particles' environment. If so, we will wish to track the changing frequency of different particle-types *and* collective-types. Following Damuth and Heisler (1988), I refer to the first approach as multi-level selection 1 (MLS1), the second as multi-level selection 2 (MLS2).

The distinction between MLS1 and MLS2 dovetails with the distinction between collective fitness$_1$ and fitness$_2$. To see this, consider an example of MLS1: D.S. Wilson's (1975) 'trait-group' model for the evolution of altruism. Organisms are of two types in this model: selfish and altruist. They assort in groups for part of their life cycle, during which fitness-affecting interactions take place, before blending into the global population and reproducing. Within each group, altruists have lower fitness than selfish types. But groups containing a high proportion

---

[17] In Sober's (1984) terminology, the issue is whether the particles or the collectives are the 'benchmarks of selection'.

of altruists have a higher group fitness$_1$, that is, contribute more *individual* offspring to the global population, than groups containing a lower proportion. So within-group selection favours selfishness, while between-group selection favours altruism; the overall outcome depends on the balance between the two selective forces. Wilson's model is thus designed to explain the changing frequency of an *individual* trait—altruism—in the overall population. Although the explanation makes essential appeal to group structure, and treats groups as fitness-bearing entities, it permits no inference about the frequency of different types of group. Both levels of selection contribute to a change in a single evolutionary parameter.

By contrast, consider D. Jablonski's hypothesis that the average geographic range of late-Cretaceous mollusc species increased as a result of species selection (Jablonski 1987). Here the *explanandum* is the fact that species with large geographic ranges became more common, in a particular mollusc clade, than those with smaller ranges. The suggested explanation is that species with larger geographic ranges had greater fitness$_2$, that is, left more offspring *species*, and that geographic range was heritable. Note that this hypothesis permits no inference about the frequency of different types of *organism*, even though the species character in question—geographic range—presumably depends on organismic characters, such as motility and dispersal. Within each species, these characters can evolve by selection at the level of the individual organism, so there is a potential interplay between the two levels of selection. But the key point is that fitnesses at each level are independently defined; so selection at each level leads to a different type of evolutionary change, measured in different units. This is the hallmark of MLS2.

Note that in Jablonski's example, the collective character subject to selection—geographic range—is emergent, but in Wilson's model it is aggregate—proportion of altruists in a group. Does it follow that the MLS1/MLS2 distinction always lines up with the aggregate/emergent distinction? The answer seems to be no. Variation between collectives with respect to emergent characters could influence the number of offspring particles they leave—in which case MLS1 would operate on an emergent character. Conversely, variation between collectives with respect to aggregate characters could influence the number of offspring collectives they leave—in which case MLS2 would operate on an aggregate character. This suggests that the MLS1/MLS2 distinction crosscuts the aggregate/emergent distinction; the issue is probed further in Chapter 4.

One might think that the MLS1/MLS2 distinction dovetails with the distinction between collectives that persist for one or a few particle generations, and those that persist for many. But this is not quite right. It is certainly most natural to define collective fitness as fitness$_2$ when particle and collective generations are non-synchronized, as in the species selection example, and as fitness$_1$ in cases where generations are synchronized, as in the trait-group example; but naturalness is not logical necessity. Even if a collective persists for many particle generations, one could still define collective fitness as average particle fitness, that is, in the MLS1 way. Conversely, even if generations are synchronized, one could still define a collective's fitness as number of offspring collectives, that is, in the MLS2 way. The essence of the MLS1/MLS2 distinction concerns the units whose demography we wish to track, which is orthogonal to the issue of their respective generation times.

One important difference between MLS1 and MLS2 is this. In MLS2, collective fitness is defined as number of offspring collectives. For this notion to apply, it is essential that the collectives reproduce in the ordinary sense, that is, that they 'make more' collectives; otherwise, determinate parent–offspring relations at the collective level will not exist. But in MLS1 this is inessential. The role of the collectives in MLS1 is to generate population structure for the particles, which affects their fitnesses. For MLS1 to produce sustained evolutionary consequences, collectives must 'reappear' regularly down the generations, but there is no reason why the collectives themselves must stand in parent–offspring relations. This is significant because, as the slime mould example illustrates, where collectives are formed by aggregation of many particles of different ancestry, the notion of collective reproduction in the sense of collectives 'making more' collectives becomes strained. In such cases it is difficult to apply MLS2 concepts, but straightforward to apply MLS1 concepts.

There is a recurring tendency in the literature to argue that MLS1 does not constitute 'real' multi-level selection at all (Maynard Smith 1976; Vrba 1989). Proponents of this view often claim that MLS1 is just lower-level selection with frequency-dependent fitnesses, so involves only one level of selection. Two points about this view deserve mention. First, it has the unwelcome consequence that most models of group selection do not deal with real group selection at all. For as Damuth and Heisler (1988) note, such models have typically been of the MLS1 variety, though they are often informally glossed using MLS2 language. This consideration is not decisive, but it does suggest that MLS1 and MLS2 should both be classified as multi-level selection, albeit of different logical types.

Secondly, although MLS1 treats the particles as the focal units, it can nonetheless shed light on collective-level phenomena. For since collectives are composed of particles, explaining the evolution of a particle character could help explain salient features of collectives too. This is especially so given that MLS1 models often focus on the particles' social behaviour, for example, their tendency to behave altruistically towards other particles, or to aggregate with them in collectives, or to police their selfish tendencies. A theory that explained how particles evolved such traits could be the first step in explaining the existence of cohesive collectives, whose constituent particles work for the good of the whole. So although the explanatory target of an MLS1 model is the change in frequency of a particle character, not a collective character, this does not mean that MLS1 can tell us nothing about the collectives themselves.

Describing MLS1 and MLS2 as different 'approaches' to multi-level selection may invite a conventionalist interpretation—as if the choice between them were a matter of taste. But this interpretation should be resisted. In a single-level scenario, whether a given trait evolves by natural selection is a matter of objective fact. Of course, our explanatory interests determine which traits we are interested in, and thus the sorts of evolutionary model we construct, but that is a different matter. The same is true of multi-level selection. MLS1 and MLS2 are distinct processes that can occur in nature; whether either occurs in a particular case is a matter of objective fact. But our explanatory interests may determine which process we wish to model, and thus which definition of collective fitness we choose.[18] Any conventionalism here is of the innocuous sort that arises because all scientific investigations must focus on some aspects of nature at the expense of others.

2.2.5 Particle Heritability and Collective Heritability

In Chapter 1, we saw that selection only produces an evolutionary response if the character subject to selection, that is, that covaries with fitness, is heritable. Heritability means parent–offspring resemblance; it is measured by the regression of offspring on parent character. In principle it should be straightforward to extend the notion of heritability

---

[18] As Damuth and Heisler (1988) say, 'once one has decided to analyse a given situation in terms of multilevel selection processes both approaches are legitimate . . . a choice has to be made depending upon what questions are of interest' (p. 411).

to a multi-level scenario. Collective heritability should mean resemblance between parent and offspring collectives; particle heritability should mean resemblance between parent and offspring particles. This is basically correct, but there are certain complications.

One complication arises from the distinction between MLS1 and MLS2. In their original presentation of this distinction, Damuth and Heisler (1988) said nothing about heritability; their focus was on selection itself, not the evolutionary response to selection. (The same is true of the other authors who have discussed the MLS1/MLS2 contrast.)[19] But the general point that heritability is necessary for a response to selection clearly must apply to multi-level selection. So the question arises: how should the concept of heritability be understood in MLS1 and MLS2? Are the relevant concept(s) of heritability the same in both cases, or not?

Consider MLS2 first. Selection at the collective level, in MLS2, means some collectives producing more offspring collectives than others. For this to produce an evolutionary response, offspring collectives must tend to resemble their parents with respect to whatever collective character is subject to selection. (In Jablonski's example, if mollusc species with large geographic ranges did not tend to give rise to daughter species with large ranges, then species selection could not lead the mean geographic range to increase.)[20] The resemblance between parent collectives and their offspring collectives can be called 'collective heritability$_2$', to parallel collective fitness$_2$. In principle, collective heritability$_2$ is straightforward to determine, by the standard offspring–parent regression technique.

What about particle heritability? In MLS2, particle-level selection operates within each collective, leading to changes in the distribution of particle characters, and thus potentially affecting the collective character too. In Jablonski's example, organismic selection can operate within each mollusc species, altering the distribution of organismic characters, and thus indirectly affecting the species' range. This means that we need to calculate a *separate* particle heritability for each collective. Look at all the particles within a single collective, consider their offspring, and calculate

---

[19] Okasha (2003b) attempts to redress this situation, by explicitly discussing heritability in relation to the MLS1/MLS2 distinction; but the analysis in that paper is flawed in several respects.

[20] Jablonski writes: 'selection at the species level could not occur unless closely related species tend to have geographic ranges more similar in magnitude that expected by chance alone' (1987 p. 361). He should really say that this is required for an evolutionary *response* to selection, not for selection itself.

the offspring–parent regression for the character of interest; then repeat the procedure for the next collective. Clearly, this will result in as many particle heritability coefficients as there are collectives. Let us call this concept '*local* particle heritability', since it measures the resemblance between parent and offspring particles within one collective.

A different possibility is to calculate a *single* particle heritability for the global population of particles. To do this, consider all the particles without regard to collective membership; look at all the offspring particles they produce, and take the offspring–parent regression. This amounts to ignoring the collectives altogether for the purposes of computing particle heritability. I call this '*global* particle heritability'. Note that the global particle heritability is *not* simply the average of the local particle heritabilities; this equality holds only as a special case.

To summarize so far. In multi-level selection of the MLS2 variety, the concept of heritability presents no great difficulties. For collective-level selection to produce an evolutionary response, collective heritability$_2$ is essential; for particle-level selection to produce an evolutionary response, *local* particle heritability is essential, given that a separate process of particle-level selection takes place within each collective.

In MLS1, by contrast, it is less obvious which are the relevant notion(s) of heritability; this point has caused confusion in the literature, as we shall see in Chapter 6. Since selection at both levels affects the evolution of a *particle* character in MLS1, intuitively it seems that collective heritability is irrelevant; what matters is parent–offspring resemblance between the particles themselves. This ties in with the point that collective reproduction in the sense of collectives 'making more' collectives is irrelevant in MLS1.

On the other hand, *some* sort of collective heritability seems necessary in MLS1. For although particles are the focal units, a selection process at the collective level does occur—the collectives make differential contributions of particles to the next generation. This will only cause evolutionary change if the set of particles produced by a collective has a similar composition to the collective itself; and this similarity is surely heritability of a sort. (In Wilson's trait-group model, if predominantly altruistic groups do not give rise to predominantly altruistic individuals, then selection between groups will not alter the overall frequency of altruists.) So *some* sort of collective heritability seems necessary in MLS1, even though the particles are the focal units.

Thus we have two conflicting intuitive arguments concerning the role of heritability in MLS1. On the one hand, collective heritability in the sense of a resemblance between parent and offspring collectives seems inessential, given that the collectives need not even form parent–offspring lineages. On the other hand, *some* sort of collective heritability seems necessary, given that a selection process at the collective level does occur. This issue can only be resolved via a mathematical description of multi-level selection, a task to which we turn next.

2.3 PRICE'S EQUATION IN A HIERARCHICAL SETTING

In Chapter 1, we used Price's equation to help understand evolution by natural selection at a single level. It is natural to hope that multi-level selection can be similarly illuminated. Price himself applied his formalism to selection at the group level, and Hamilton (1975) showed how the formalism can be extended to a general multi-level scenario, with an indefinite number of hierarchical levels. As noted earlier, the Price formalism is now routinely used by theorists interested in multi-level selection. This section outlines the Price approach to multi-level selection and draws some morals.

2.3.1 The Price Approach to MLS1

Recall the first version of Price's equation derived in Chapter 1:

$$\overline{w}\Delta\overline{z} = \text{Cov}(w_i, z_i) + E(w_i \Delta z_i) \qquad (2.1)$$

where $\Delta\overline{z}$ is the change in average character from one generation to another; w_i is the absolute fitness of the ith entity; $\overline{w}$ is average fitness; z_i is the character value of the ith entity, and Δz_i is the deviation in character between the ith entity and the average of its offspring.[21] Recall that Cov (w_i, z_i) is the character-fitness covariance, measuring the extent to which the character z is subject to selection, while $E(w_i \Delta z_i)$ is the average of the parent–offspring deviations multiplied by fitness.

[21] Recall that 'fitness' in this equation means *realized* fitness, i.e. drift is ignored. Deriving a multi-level version of Price's equation that takes account of drift would be straightforward, but more cumbersome.

Selection at Multiple Levels: Concepts and Methods 63

Equation (2.1) is designed for a single-level scenario, that is, a population where all entities occupy the same hierarchical level. But suppose we are dealing with a two-level scenario, where particles are nested in collectives. (For simplicity, assume that all collectives initially contain equal numbers of particles.) Equation (2.1) still holds true, provided that the index i is taken to range over all particles in the global population. Of course, this interpretation of equation (2.1) simply amounts to ignoring the collectives—characters and fitnesses are attributed only to the particles, and the equation tells us only about the evolution of the mean particle character.

However, starting from this interpretation, we can decompose the Cov term of equation (2.1) to take account of the collectives explicitly. Cov (w_i, z_i) is the *overall* covariance between particle fitness and character, in the global population of particles. Since the particles are nested within collectives, this covariance can be analysed into two components—a 'within-collective' component and a 'between-collective' component.[22] Since $\Delta \bar{z}$ depends directly on Cov (w_i, z_i), this allows us to partition $\Delta \bar{z}$ itself into within- and between-collective components. To perform the decomposition, we need some new notation.

We let w_{jk} denote the absolute fitness of the jth particle in the kth collective. Similarly, we let z_{jk} denote the character value of the jth particle in the kth collective. So 'k' indexes collectives, while 'j' indexes particles within collectives; 'i' continues to index particles in the global population, ignoring the collectives. Therefore, $\bar{z} = \frac{1}{n} \sum_i z_i = \frac{1}{n} \sum_k \sum_j z_{jk}$ where n is the number of particles in the global population. Similarly, $\bar{w} = \frac{1}{n} \sum_i w_i = \frac{1}{n} \sum_k \sum_j w_{jk}$

Since we are dealing with multi-level selection, we need to attribute fitnesses and characters to collectives as well as particles. Let W_k denote the fitness of the kth collective, defined as the average fitness of its constituent particles, that is, $W_k = \frac{1}{N} \sum_j w_{jk}$, where N is the number of particles per collective. Note that this is collective fitness$_1$. Similarly, we let Z_k denote the character value of the kth collective, defined as the average character value of its constituent particles, that is, $Z_k = \frac{1}{N} \sum_j z_{jk}$. Note that this collective character is 'aggregate'

---

[22] In precisely the same way, the total variance of a variable, in any grouped population, can be analysed into within-group and between-group components; this statistical technique is known as the 'one way' analysis of variance.

not 'emergent'. It is important to remember that W_k and Z_k denote properties of the collectives, not the particles.

Given these definitions, it immediately follows that:

$$\underset{\text{covariance}}{\text{Overall}} \qquad \underset{\text{covariance}}{\text{Collective-level}} \qquad \underset{\text{covariance}}{\text{Average particle-level}}$$

$$\text{Cov}(w_i, z_i) = \text{Cov}(W_k, Z_k) + E_k(\text{Cov}_k(w_{jk}, z_{jk})) \qquad (2.2)$$

which is the decomposition referred to above.[23] This equation tells us that the *overall* character-fitness covariance, in the global population of particles, equals the sum of two components. The first component, $\text{Cov}(W_k, Z_k)$, is the covariance between the collective means. It measures the extent to which differences in collective character Z_k are associated with differences in collective fitness W_k. The second component, $E_k(\text{Cov}_k(w_{jk}, z_{jk}))$, is the average of the within-collective covariances between particle character and particle fitness. To calculate this term, consider each collective in turn, find the covariance between w and z within the collective, then take the average across the collectives. So $\text{Cov}_k(w_{jk}, z_{jk})$ denotes the character-fitness covariance *within* the kth collective, and E_k the average across all the collectives. For simplicity, the subscripts will be dropped wherever possible, yielding the more tractable form:

$$\text{Cov}(w_i, z_i) = \text{Cov}(W, Z) + E(\text{Cov}_k(w, z))$$

Note that equation (2.2) makes no new biological assumptions, it depends only on the fact that the particles in the global population are nested within collectives.[24]

The significance of equation (2.2) is that it allows us to study the combined effects of two levels of selection on the character z. To see this, we can substitute equation (2.2) directly into the Price equation for the global population of particles (equation (2.1)). If we assume that the particles 'breed true', that is, pass on their character without deviation, then the Price equation reduces to:

$$\overline{w}\Delta\overline{z} = \text{Cov}(w_i, z_i)$$

---

[23] See Wade (1985) or Frank (1998) for a full derivation of equation (2.2).
[24] If we set z =w, i.e. let the phenotypic trait z be fitness itself, then equation (2.2) reduces to the standard formula for a one-way ANOVA.

Substituting in equation (2.2) gives:

$$\overline{w}\Delta\overline{z} = \underset{\text{selection}}{\underset{\text{Collective-level}}{\text{Cov}(W, Z)}} + \underset{\text{selection}}{\underset{\text{Particle-level}}{\text{E}(\text{Cov}_k(w, z))}} \qquad (2.3)$$

which is a multi-level version of Price's equation. This equation expresses the total change in mean particle character, $\Delta\overline{z}$, as the sum of two components. It is natural to interpret these components as corresponding to selection at the level of collectives and particles respectively, as did Price (1972) and Hamilton (1975). On this view, Cov (W, Z) represents the effect of collective-level selection on $\Delta\overline{z}$, while E $(\text{Cov}_k(w, z))$ represents the effect of particle-level selection. Hamilton (1975) expressed this by saying that the Price equation, in its hierarchically decomposed version above, effects a 'formal separation of levels of selection' (p. 333). I call this the 'Price approach' to MLS1.

The grounds for the Price approach are easy to see. Suppose that all the collectives have the same fitness, or that collective fitness W is uncorrelated with collective character Z. Intuitively there can be no collective-level selection in such a situation. And the term Cov (W, Z) will equal zero, in that situation. Conversely, suppose that within each collective, all the particles have the same fitness, or that particle fitness w is uncorrelated with particle character z. Intuitively there can be no particle-level selection in such a situation. And the term E $(\text{Cov}_k(w, z))$ will be zero in that situation—for each of the within-collective covariances will be zero, so their average will be zero.[25] So given the plausible idea that collective-level selection means selection *between* collectives, and particle-level selection means selection *between* particles within the *same* collective, equation (2.3) decomposes the total change in $\overline{z}$ into components corresponding to each level of selection.

Note that equation (2.3) describes MLS1, not MLS2: the particles constitute the focal units. Indicative of this is that *both* components contribute to a change in the mean *particle* character $\Delta\overline{z}$, and that collective fitness W is, by definition, equal to average particle fitness.[26] So collective selection, as represented in equation (2.3), means collectives making

---

[25] However, in general one cannot infer that if E $(\text{Cov}_k(w, z)) = 0$, there is no selection at the particle level. One can only infer particle-level selection has no *net* effect on the evolutionary change. This is compatible with the existence of particle-level selection, if it operates in a different direction in different collectives.

[26] Also indicative of this is the fact that, if the number of particles per collective were not initially equal, the Cov and Exp terms in equation (2.3) would need weighting by

differential contributions of *particles* to the next generation. Note also that the collective character Z is, by definition, average particle character. Of course, variation with respect to some other collective character, for example, an emergent one, could also lead to differences in collective fitness$_1$. But the only collective character that can explicitly appear in equation (2.3) is average particle character. This is a representational limitation of equation (2.3), not an indication that the collective characters causally implicated in MLS1 must be of a particular type.

In Chapter 1, we distinguished between *statistical* and *causal* decompositions of the total evolutionary change. What becomes of this distinction in relation to the multi-level version of Price's equation? Equation (2.3) above certainly provides a statistical decomposition of $\Delta \bar{z}$ in a multi-level scenario, but do the components correspond to distinct causal factors? Prima facie, the answer seems to be yes: Cov (W, Z) equals the change caused by collective-level selection, while E (Cov$_k$(w, z)) equals the change caused by particle-level selection. This was the interpretation intended by Price and Hamilton—though they did not use overtly causal language—and has been explicitly endorsed by Sober and Wilson (1998) in relation to individual and group selection. Such an interpretation seems reasonable, modulo the usual caveats about causation and correlation, but later we shall see that the issue of causal decomposition in multi-level scenarios raises unexpected subtleties.

The full power of the Price approach to MLS1 may not be evident from the abstract treatment above, so two applications of equation (2.3) are presented in the next section. The first is to the evolution of an altruistic trait in a multi-group setting, the second to the evolution of a segregation-distorter allele in a population of diploid organisms.

2.3.2 Applications

Consider a simplified version of D. S. Wilson's (1975) 'trait-group' model for the evolution of altruism, described above. Organisms of two types, altruistic (A) and selfish (S) assort in groups of size N for a period of their life cycle, where they engage in fitness-affecting interactions. They then blend into the global population, reproduce, die immediately, and the cycle repeats; see Figure 2.8. Reproduction is asexual and transmission is perfect: altruists always give rise to altruists, and similarly for selfish types.

collective size, as both Price (1972) and Hamilton (1975) noted. This highlights the fact that the particles are being counted, not the collectives, as Damuth and Heisler (1988) stress.

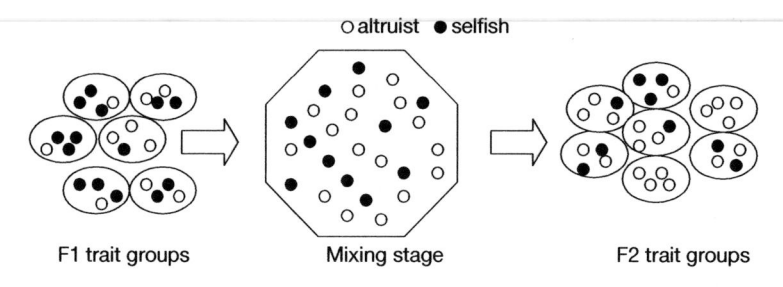

Figure 2.8. Wilson's trait-group model for the evolution of altruism

Altruists engage in group-beneficial behaviours; selfish types do not. So the greater the proportion of altruists in a group, the higher the group's fitness$_1$, that is, the average fitness of its members. But within any mixed group, a selfish type is fitter than an altruist—for the former receives benefits from the latter but not vice versa. So selection between individuals within groups favours selfishness; but selection between groups favours altruism. Depending on the balance between the two selective forces, altruism may be able to spread in the global population.

These ideas are easily expressed in terms of the Price formalism. The individual organisms are the particles and the trait groups are the collectives. We define $z = 1$ for altruists, $z = 0$ for selfish types. So $\bar{z}$ equals the overall proportion of altruists in the global population. The variables W and Z have their obvious interpretation: a group's W-value is the average fitness of its constituent organisms, while its Z-value is the proportion of altruists it contains. Since organisms are assumed to breed true, transmission bias is zero, so equation (2.3) applies. It tells us that $\Delta\bar{z}$, the change in the overall proportion of altruists from one generation to another, is affected by two components. The Cov (W, Z) component reflects the impact of group selection on $\Delta\bar{z}$, while the E (Cov$_k$ (w, z)) component reflects the impact of individual selection.

This illustrates two important points. First, the *direction* of selection may be different at the two levels. The term Cov (W, Z) must be positive—the greater the proportion of altruists in a group, the higher the group's fitness. By contrast, E (Cov$_k$ (w, z)) must be negative—within each (mixed) group there is a negative correlation between fitness and altruism, so the average across all the groups must be negative too. This immediately gives us the condition required for altruism to evolve: Cov (W, Z) must be greater than the absolute value of E (Cov$_k$ (w, z)). If Cov (W, Z) > |E (Cov$_k$(w, z))| then $\Delta\bar{z} > 0$, so altruism spreads;

if Cov (W, Z) < |E (Cov$_k$(w, z))| then $\Delta \bar{z}$ <0 so selfishness spreads; while if Cov (W, Z) = |E (Cov$_k$(w, z))| then $\Delta \bar{z}$ = 0, so there is no evolution, that is, group and individual selection exactly cancel out each other's effect. Interestingly, this condition for the spread of altruism is closely related to the famous Hamilton rule (cf. Goodnight et al. (1992), Frank (1998), Queller (1992b), Okasha (2005)).

Secondly, the balance between the two levels of selection is affected by *positive assortment*. Selection at any level requires variance at that level: if all units have the same character, or the same fitness, no selection is possible. As equation (2.3) illustrates, group selection requires variation between groups, with respect to both fitness and character, while individual selection requires variation within groups. So any mechanisms which increase the between-group variance, or reduce the within-group variance, will enhance the power of group selection relative to individual selection. One such mechanism is positive assortment, that is, like assorting with like. If altruists tend to form groups with other altruists and selfish types do likewise, this will make the groups more internally homogenous, reducing the scope for within-group selection but increasing the scope for between-group selection. This general principle applies at all hierarchical levels.

Our second illustration of the multi-level Price equation at work is quite different. Consider a simple diploid population-genetic model, with two alleles, A and B, segregating at a locus. So there are three genotypes in the population, AA, AB, and AA, which can be thought of as groups of two alleles (Figure 2.9). The multi-level concepts then apply neatly: the particles are individual alleles, the collectives are diploid organisms. As discussed in Section 2.2.2, this means that particle-level selection occurs when the two alleles within an AB heterozygote differ in fitness, that is, when meiotic drive occurs; while collective-level selection occurs when there are fitness differences between the genotypes themselves. The change in allele frequency from one generation to another depends on both selective forces.

This model is easily described in terms of the Price formalism. We define $z = 1$ for an A allele, $z = 0$ for a B allele. An allele's w-value is its fitness, that is, the number of copies it bequeaths to the (successful) gamete pool.[27] Z and W have their obvious interpretation.

---

[27] Note that 'allelic fitness' as the term is used here means the fitness of an allele within a diploid genotype. This is *not* the same as the 'marginal allelic fitness' of standard population genetics textbooks, which is the fitness of an allele-type averaged across all

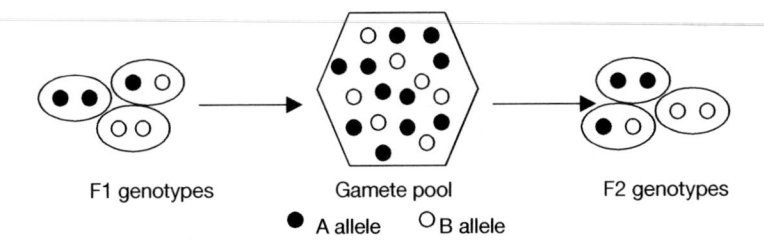

Figure 2.9. Multi-level approach to diploid population genetics

An organism's Z-value is the proportion of A alleles it contains, which equals 1 for AA, $1/2$ for AB, and 0 for BB; an organism's W-value is the average fitness of the two alleles it contains, that is, half its total gametic contribution to the next generation. $\bar{z}$ is the overall frequency of the A allele in the population, and $\Delta\bar{z}$ is the change in frequency from one generation to another. Formally at least, this is exactly isomorphic to the group selection example.

It may not be immediately clear why diploid population genetics with meiotic drive counts as an instance of MLS1, rather than MLS2. When the number of particles per collective is constant, the difference between the two sorts of multi-level selection tends to be obscured, for counting weighted or unweighted gives the same answer. This applies here—the number of alleles per organism is fixed, so it is unclear whether $\bar{z}$ should be thought of as the average value of the allelic character z, or the organismic character Z. However, suppose that the ploidy level was variable. Then, the average allelic character would only equal the average organismic character if the latter were weighted by ploidy, that is, collective size. I take it that in this case, the ploidy-weighted measure is the natural one, which means that we are counting particles not collectives, hence dealing with MLS1.[28]

To apply equation (2.3), we assume that the particles breed true, that is, there is no mutation. The equation then tells us that the change in frequency of the A allele, $\Delta\bar{z}$, depends on Cov (W, Z), which reflects selection between genotypes, and E (Cov$_k$ (w, z)), which reflects selection within genotypes—or what is sometimes called 'genic

---

genotypes. See Chapter 5, Section 5.3 and Kerr and Godfrey-Smith (2002) for further discussion of this point.

[28] The point that the Cov term of Price's equation must be weighted by ploidy, if ploidy levels are variable, has been discussed by Grafen (1985) and Seger (1981).

selection'.[29] If segregation is perfectly Mendelian, that is, if there is no meiotic drive, then E (Cov$_k$ (w, z)) must be zero. For within the AA and BB homozygotes, there can obviously be no covariance between w and z, since neither varies; and Mendelian segregation means that the two alleles in an AB heterozygote have identical fitness. So E (Cov$_k$ (w, z)) measures the impact of meiotic drive on the allele frequency change.

The heuristic utility of viewing diploid population genetics this way is clearest when the direction of selection is different at the two levels. Suppose the A allele distorts segregation in its favour. This means that E (Cov$_k$ (w, z)) is positive. But suppose the A allele also has deleterious effects on organismic fitness—as is usually the case for segregation-distorters. This means that AA organisms suffer a fitness disadvantage vis-à-vis BB organisms; so there is a negative correlation between organismic fitness W and character Z. Therefore, Cov (W, Z) is negative. The overall evolutionary outcome then depends on which selective force is stronger, just as in the group selection example.

This example illustrates an important point which Price himself noted: selection at one hierarchical level appears as transmission bias at a higher level. Since we stipulated that there is no mutation, transmission bias at the *particle* level is impossible: each particle transmits its character faithfully. However, a sort of transmission bias at the *collective* level is possible. A collective may imperfectly transmit its collective character to the set of offspring particles that it produces. For example, an AB heterozygote, whose character value is $Z = 1/2$, may give rise to a set of offspring alleles in which the frequency of the A allele is greater or less than $1/2$. Clearly, this is simply the possibility of segregation-distortion itself. So we can think of the E (Cov$_k$ (w, z)) term as reflecting either selection between the particles, or the imperfect transmission of a collective character; these amount to the same thing. This point is sometimes expressed by saying that selection at one level reduces the fidelity of transmission at higher levels (Michod 1999).

The same point holds in the group selection example. As we described that example, E (Cov$_k$ (w, z)) reflects within-group selection—which occurs because selfish organisms in a mixed group have higher fitness than altruists. But we can also think of this as the imperfect transmission

---

[29] Note however that the expression 'genic selection' has been used in other ways too; see Chapter 5.

Selection at Multiple Levels: Concepts and Methods 71

of a group-level character. Imagine a group with character value Z = ½, that is, a group containing 50 per cent altruists and 50 per cent selfish organisms. The set of offspring organisms produced by this group will contain more than 50 per cent selfish, owing to within-group selection. So the group has imperfectly transmitted its character value to this set of offspring organisms. The fact that selection at one level equates to transmission bias at a higher level is explained by the *recursive* nature of the Price equation (cf. Hamilton (1975), Frank (1998)). The recursiveness is evident from equation (2.3): the expression inside the 'Exp' term has the same form as the 'Cov' term itself.

2.3.3 Heritability in MLS1 Revisited

The issue of heritability in MLS1, left pending in Section 2.2.5, can now be resolved. In Chapter 1, we derived a single-level version of Price's equation in which heritability features explicitly. First we showed that the total response to selection equals Cov (ω_i, z'_i)—the covariance between an entity's relative fitness and the average character of its *offspring* z'. Then we showed that, modulo an important assumption about the error covariance, Cov (ω_i, z'_i) can be written as the product of the selection differential Cov (ω_i, z_i) and the heritability h. In this section we deal with absolute rather than relative fitness, so we shall extend the meanings of 'selection differential' and 'response to selection' slightly, taking them to refer to Cov (w_i, z_i) and Cov (w_i, z'_i) respectively, rather than to their relative fitness counterparts.[30] So we can write:

Response to selection   Heritability x selection differential

$$\text{Cov}(w_i, z'_i) \quad = \quad h\,\text{Cov}(w_i, z_i)$$

In a multi-level scenario the term Cov (w_i, z_i) may be partitioned into within- and between-collective components, given by equation (2.3). Substituting into the above equation, this gives:

Total response to selection   Heritability x (collective selection differential
                                           +average particle selection differential)

$$\text{Cov}(w_i, z'_i) \quad = \quad h\,(\text{Cov}(W, Z) + E\,(\text{Cov}_k(w, z))) \qquad (2.4)$$

---

[30] This terminological extension is innocent; to eliminate it, simply divide each of the equations in this section by average fitness $\bar{w}$.

This suggests that *global* particle heritability is the relevant type of heritability in MLS1, for this is what 'h' denotes. Recall that global particle heritability is the parent–offspring resemblance in the global population of particles. So although (2.4) is a multi-level equation, in which collective fitness W and collective character Z explicitly appear, the only *heritability* term that explicitly appears is (global) particle heritability. This should not be surprising: it tallies with the point that collective reproduction in the sense of collectives making more *collectives* is irrelevant to MLS1.

However, we can also derive a slightly different version of the multi-level Price equation, which also describes MLS1, but in which *two* heritability terms appear, one for the particles and one for the collectives (cf. Queller 1992b). To see how this works, note that we can easily partition the term Cov (w_i, z'_i) into within- and between-collective components, just as we did for Cov (w_i, z_i). This gives:

Total response to selection	Response to collective selection	Response to particle selection

$$\text{Cov}(w_i, z'_i) = \text{Cov}(W, Z') + E(\text{Cov}_k(w, z')) \qquad (2.5)$$

In equation (2.5), Cov (W, Z') is the covariance between a collective's fitness and the average z'-value of its constituent particles; E (Cov$_k$(w, z')) is the average of the within-collective covariances between particle fitness and particle z'-value. So equation (2.5) partitions the total *response to selection* into within- and between-collective components; contrast this with equation (2.3), which partitions the total *selection differential* into within- and between-collective components.

The significance of this is that any response-to-selection term can be written as the product of a selection differential and a heritability (again, modulo the assumption about the error covariance). So we can write:

Response to collective selection	Collective heritability	Collective selection differential

$$\text{Cov}(W, Z') = H_c \times \text{Cov}(W, Z)$$

where H_c is the collective heritability. Importantly, H_c is *not* collective heritability$_2$ as defined above—it does not measure the resemblance between parent and offspring collectives. Rather, it measures the resemblance between a parent collective and the set of offspring *particles* it produces.[31] Therefore, H_c measures the fidelity with which collectives transmit their average particle character to the set of offspring particles they produce.[32] Let us call H_c collective heritability$_1$, to parallel collective fitness$_1$.

Similarly, each of the $Cov_k(w, z')$ terms, which measure the response to particle selection within the kth collective, can be written as the product of a selection differential and a heritability:

Response to particle selection		Local particle heritability		Particle selection differential
$Cov_k(w, z')$	=	$h_{p(k)}$	×	$Cov_k(w, z)$

where $h_{p(k)}$ is the particle heritability within the kth collective. Note that $h_{p(k)}$ is the *local* particle heritability of z as defined previously, that is, it measures the resemblance between the parental particles in the kth collective and their offspring particles. So each of the collectives has its own $h_{p(k)}$ coefficient.

This means that equation (2.5) can be rewritten as:

Total response to selection		Response to collective selection		Response to particle selection
$Cov(w_i, z'_i)$	=	$H_c Cov(W, Z)$	+	$E(h_{p(k)} Cov_k(w, z))$

which is a multi-level equation in which *two* heritability terms feature, H_c, or collective heritability$_1$, and $h_{p(k)}$, or local particle heritability. Equation (2.5) is mathematically equivalent to equation (2.4), in which only *one* heritability term features, the global particle heritability h.[33]

This shows that MLS1 can be thought about in two different, equally legitimate ways. Either we can regard the two levels of selection as contributing to the overall selection differential on the particle character

---

[31] To see this, note that H_c is the linear regression of Z' on Z. Recall that a collective's Z-value is the average z-value of its constituent particles, while its Z'-value is the average z'-value of its constituent particles; and a particle's z'-value is the average z-value of its offspring.

[32] The quantity known as the 'heritability of the family mean' in quantitative genetics is a heritability coefficient of this sort.

[33] By combining equations (2.4) and (2.5), it is straightforward to derive an expression for the global particle heritability in terms of the collective heritability$_1$ and the local particle heritabilities; the expression is somewhat cumbersome, however.

z, which is then converted into the cross-generation evolutionary change via the global particle heritability. Alternatively, we can regard the two levels of selection as contributing to the total response to selection, where the contribution of each level depends on the selection differential *and* the heritability at that level. These two conceptions are mathematically equivalent.

2.3.4 The Price Approach to MLS2

Applying the Price formalism to MLS2 is straightforward. As before, we let Z_k denote the character value of the kth collective. But now Z may be aggregate or emergent—it is not necessarily the average of any particle character, though it may be. We let Y_k denote the absolute fitness of the kth collective. Since we are dealing with MLS2, this is collective fitness$_2$, that is, number of offspring *collectives*. So Y_k is *not* defined as the average fitness of the particles in the kth collective; it bears no necessary relation to the latter quantity. Applying the simple Price equation, it follows that:

$$\overline{Y}\Delta\overline{Z} = \text{Cov}(Y_k, Z_k) + \text{E}(Y_k \Delta Z_k) \qquad (2.6)$$

which expresses the total change in average collective character as the sum of two components, in the usual way.

Equation (2.6) is just a single-level Price equation in which the focal units are the collectives; the particles get no explicit mention. As discussed, it is likely that Z and Y will bear *some* relation to underlying particle characters, albeit perhaps an indirect one. So it is possible that particle-level selection, acting within each collective, may affect a collective's character value Z_k, or its fitness Y_k, or its transmission fidelity ΔZ_k. However, within-collective selection of this sort is not explicitly representable in equation (2.6). This is a reflection of the point stressed earlier, that in MLS2 the two levels of selection lead to different types of evolutionary change, measured in different units.

Of course, selection at the level of particles can be separately described by a Price equation. Suppose that within each collective, the particle character z is undergoing selection. We can then write:

$$\overline{w}\Delta\overline{z}_k = \text{Cov}(w, z) + \text{E}(w\Delta z) \qquad (2.7)$$

where $\overline{z}_k$ is the average particle character within the kth collective. Note that equation (2.7) applies to each collective separately, describing an evolutionary process taking place among the particles within it. Conceptually it is simplest to think of this process as occurring within the lifespan of the collective, which means that particles have a shorter generation time than collectives, so a collective's $\bar{z}_k$ value changes as it ages (or develops). But this interpretation is not strictly necessary; equations (2.6) and (2.7) say nothing about the timescales over which Y and w are measured.

We saw that in MLS1, particle-level selection *automatically* results in transmission bias at the collective level. The same is not true in MLS2. Particle-level selection within the kth collective *may* affect the collective's transmission bias ΔZ_k, but not necessarily. For example, somatic selection between the cells within a multicelled organism *may* affect the fidelity with which the organism transmits its phenotypic character to its offspring, but not necessarily. This difference between MLS1 and MLS2 reflects the fact that 'collective transmission bias' means something different in each. In MLS1, it is the character deviation between a collective and the set of offspring *particles* it produces; since collective character is defined as mean particle character, this deviation is necessarily affected by particle-level selection. In MLS2, it is the character deviation between a collective and its offspring collectives; this deviation bears at most a contingent relation to any particle-level processes, including selection. Again, this is indicative of the fact that in MLS2, the evolutionary changes caused by the two levels of selection are incommensurable.

Finally, note that the notion of heritability presents no particular difficulties in MLS2. Equations (2.6) and (2.7) could both be rewritten in terms of heritability, using the technique employed in the previous section, as can any Price equation; but no additional clarification would ensue. It is clear that in (2.7), the relevant heritability coefficient is collective heritability$_2$, and in (2.6) it is the *local* particle heritability. Collective heritability$_1$ and global particle heritability play no role in MLS2; they pertain solely to MLS1.

3

Causality and Multi-Level Selection

INTRODUCTION

The previous chapter looked at the concepts involved in multi-level selection. This chapter also deals with multi-selection, but from a different angle. The focus here is on causality. Section 3.1 argues that the levels-of-selection question is ultimately a question about the hierarchical level(s) at which there is a causal, rather than just a statistical, link between character and fitness. This raises an important possibility: a character-fitness covariance at one level may be a side effect, or by-product, of selection acting at a *different* level. The remaining sections examine such cross-level by-products in relation to both MLS1 and MLS2, and explore their consequences.

3.1 CAUSES, CORRELATIONS, AND CROSS-LEVEL BY-PRODUCTS

We noted in Chapter 1 that Darwinian explanations are causal. To attribute the evolution of a character to natural selection is to say that differences in the character caused differences in fitness, and that the character was heritable. In a single-level scenario, where all entities are at the same hierarchical level, the concept of causality presents no great difficulties, or at least none that are unique to evolutionary theory. Of course, it may be difficult to *find out* whether a given character causally influences fitness,[1] or did so in the evolutionary past, but this is strictly an epistemological problem. It is perfectly clear what it *means* for

---

[1] To say that a character 'causally influences fitness' means that *differences* in the character cause fitness *differences*. The latter formulation is more accurate, but the former is often stylistically preferable.

character differences to cause fitness differences, however hard it may be to discover such causal relations in nature.[2]

In a multi-level scenario matters are less simple. Presumably, multi-level selection involves causality at more than one level of the hierarchy. But this raises a number of questions. Are the higher- and lower-level causal processes autonomous, or are they interdependent? Might selection at one level ever be 'reduced' to selection at a lower level? If higher-level selection has an impact on lower-level phenomena, does this mean that 'downward causation' is occurring? (The significance of this question is that some philosophers regard downward causation as a suspect notion.) Interestingly, parallel questions arise in the philosophy of psychology about the relation between mind and brain, and in the philosophy of social science about the relation between individual and society. Whether or not these parallels can be fruitfully exploited, their mere existence suggests that the concept of causality, as it relates to multi-level selection, deserves close scrutiny.

The Price equation provides a useful starting point. It tells us that multi-level selection (whether of the MLS1 or MLS2 variety) requires character-fitness covariance at more than one level. Covariance is of course a statistical notion, not a causal one. If two variables are related as cause and effect they will normally covary, but the converse inference, from covariance to causality, is fraught with danger, as everyone knows. This point was discussed in relation to the single-level Price equation in Chapter 1; it applies equally to the multi-level expansions examined in Chapter 2.

In the single-level case, we captured the point about covariance and causality by distinguishing 'direct' from 'indirect' selection, depending on whether the character in question causally influences fitness, or correlates with it for some other reason. This distinction extrapolates smoothly to the multi-level case. In a two-level scenario, where particles are nested within collectives, a given particle character may covary with particle fitness because it causally influences it, or for some other reason; similarly for a collective character and fitness. So the direct/indirect distinction applies at both levels.

Conceptualizing the issue this way raises an important possibility, namely that a character-fitness covariance at one level, for example, the collective level, may be a side effect of direct selection at a *different*

---

[2] See Endler (1986) for a comprehensive analysis of the empirical methods that can help determine whether a given character causally affects fitness.

level, for example, the particle level. That this is possible seems clear. For properties of collectives normally depend on properties of their constituent particles; so a causal process occurring at the particle level, such as natural selection, may have effects that 'filter up' the hierarchy, producing a covariance between two collective properties, character and fitness, that are not themselves causally related. If so, the character-fitness covariance at the collective level will be a *by-product* of lower-level selection.

By-products running up the hierarchy, from the particle to collective level, are easily envisaged. But what about the reverse direction? Could a character-fitness covariance at the particle level be a by-product of selection acting at the collective level? This possibility is less easy to envisage but should not be ruled out a priori; we return to it below.

Cross-level by-products obviously complicate the task of understanding the causal basis of evolutionary change. In a single-level scenario, it may be difficult to determine whether a given character is being directly selected, but there is no room for error about the *level* at which selection is occurring. In a multi-level scenario both types of error are possible. Not only is it possible to be wrong about which character(s) are being directly selected, it is also possible to be wrong about which level the causal process of selection is occurring at, owing to the possibility of a cross-level by-product.

However, the problem is not merely epistemological; it cuts deeper than that. For once cross-level by-products are in the picture, it ceases to be entirely clear what it *means* for there to be a direct causal link between character and fitness at a given level. Since there are always likely to be causal arrows running from lower to higher levels, and possibly in the converse direction, the requirements that must be satisfied for a given character-fitness covariance to reflect causality at the level in question are not obvious. Some sort of 'autonomy' from causal processes at other levels is presumably required, but this notion needs elucidation. In just the same way, the parallel problems in philosophy of psychology about mind and brain, and in philosophy of social science about individual and society, are not purely epistemological. In each case, the problem is not just to discover whether phenomena at a given level *are* by-products of causal processes at other levels, but also to determine precisely what this amounts to.

This suggests a way of formulating the levels-of-selection question that is particularly sharp. The question becomes: *when is a character-fitness covariance indicative of direct selection at the level in question, and when is it a by-product of direct selection at another hierarchical level?* I suggest that this is the question actually at stake in debates over the levels of selection, though they are rarely formulated in precisely these terms.

This formulation of the question has three advantages. First, it provides a unified framework for understanding a range of specific debates. The literature is replete with claims that biological phenomena at one level are by-products of selection acting at a different, usually lower, level. For example, G. C. Williams (1966) stresses the difference between genuine group adaptations and 'fortuitous group benefits', which are side effects of selection on individuals. Similarly, Vrba (1989) and Eldredge (1989) argue that the macroevolutionary trends often attributed to species selection are really by-products of selection at lower hierarchical levels. More recently, in their work on the evolution of multicellularity, Michod and Roze (1999) argue that a crucial stage occurred when the fitness of the emerging multicelled organism became 'decoupled' from the fitnesses of its constituent cells; before this stage, any differences in organismic fitness were a side effect of selection at the cellular level. Our formulation captures a common thread running through these disparate discussions.

Secondly, our formulation helps make sense of various proposals in the literature for how to determine the 'real' level(s) of selection. For example, Hull's (1981) idea that selection at any level requires entities at that level to 'interact as cohesive wholes with their environment' can be read as an informal specification of the conditions required for a given character-fitness covariance to be interpreted in causal terms. Similarly, the 'additivity criterion' of Wimsatt (1980) and Lloyd (1988) can be seen as an attempt to decide whether a given character-fitness covariance reflects direct selection at the level in question, or whether it is the result of selection at another level. The same is true of Vrba's (1989) 'emergent character' criterion, Gould's (2002) 'emergent fitness' criterion, and Brandon's (1990) 'screening off' criterion. This enables us to compare and assess these competing proposals, a task undertaken in Chapter 4.

Thirdly, our formulation helps explain why the levels-of-selection debate is in part philosophical. The Price analysis developed previously, for all the clarification it brings, gives no indication of why disagreements about the level of selection can be anything other than empirical. The obvious inference to draw from Price is that to determine the level(s) at which selection is acting, in any given case, requires discovering the level(s) at which fitness and character covary, which is an empirical task. Of course, this view needs qualifying by the standard caveats about correlation not implying causation, and the consequent need to distinguish direct from indirect selection. But it needs qualifying further, and more fundamentally, by the possibility of cross-level by-products; for they throw into question the very idea of a 'direct' causal relation between character and fitness. It is here that the debate ceases to be purely empirical, and philosophical conceptions intrude.

3.2 SELECTION ON CORRELATED CHARACTERS

It helps to think abstractly about why two variables might correlate in the absence of a causal link between them. One obvious reason is if they are joint effects of a common cause. Suppose that alcohol consumption causes both obesity and kidney failure. These two conditions will correlate, even though they are causally independent: obesity does not cause kidney failure, and kidney failure does not cause obesity. This familiar point is illustrated in Figure 3.1, where the thick arrows indicate causal influence and the dotted line indicates correlation. Statisticians refer to correlations between causally independent variables as 'spurious'.

A slight variant of the common cause scenario arises if one of the causal links is converted into a correlation but the other is left intact. Continue to suppose that alcohol consumption causes kidney failure. But now suppose alcohol does *not* cause obesity, but correlates closely with it for some other reason. Clearly, the correlation between obesity and kidney failure will be unchanged. So either of the thick arrows in Figure 3.1 can be converted into a dotted line without affecting the dotted line on the horizontal.

obesity .... kidney failure

alcohol consumption

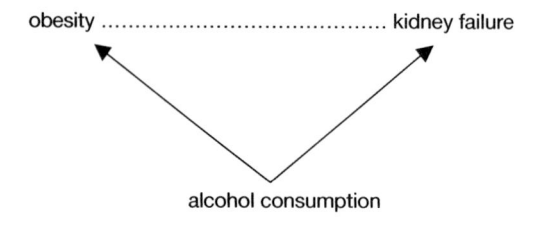

Figure 3.1. Joint effects of a common cause

Causality and Multi-Level Selection

Turning to natural selection, consider firstly a single-level scenario. Let w denote fitness, and z a phenotypic character. One way in which Cov (w, z) could be non-zero, in the absence of a causal link between z and w, is if z correlates with another character y which *does* affect fitness. For example, suppose that body size (z) and brain size (y) are positively correlated in a given population.[3] Having a large brain boosts an organism's fitness, but body size has no effect on fitness. This will result in a spurious correlation between body size and fitness, illustrated in Figure 3.2. Large-bodied organisms will be fitter than average not *because* they are large-bodied, but rather because they are large-brained and body and brain size are correlated.

```
fitness (w)------------------------------- body size (z)
     ▲                                    ╱
      ╲                                  ╱
       ╲                                ╱
        ╲                              ╱
         ╲                            ╱
          ╲                          ╱
                 brain size (y)
```

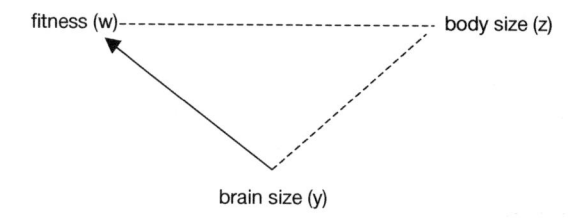

Figure 3.2. Allometry as a source of spurious character-fitness covariance

Since we are assuming a single-level scenario, that is, ignoring hierarchical structure, this example raises no levels-of-selection issues. But it is worth looking at the standard analytical technique for dealing with selection on correlated characters, as it will prove useful later. The technique is based on multiple regression analysis; it dates back to Pearson (1903) but has been most thoroughly developed by Lande and Arnold (1983).

Continuing with the same notation, assume for simplicity that z and y are the *only* two characters that causally affect fitness.[4] Then, we can capture the dependence of fitness on both characters with a standard linear regression model:

$$w = \beta_1 z + \beta_2 y + e \qquad (3.1)$$

where fitness is the response variable, z and y are the two independent variables, and e is the residual. β_1 is the partial regression of w on z,

---

[3] Such correlations are known as 'allometric'.
[4] Alternatively, we could assume that any other characters that do causally affect fitness are uncorrelated with both z and y.

controlling for y; β_2 is the partial regression of w on y, controlling for z. So β_1 tells us how much the fitness of an organism will change, on average, when its z-value is increased by one unit *while its y-value is held fixed*. Similarly, β_2 tells us about how much an organism's fitness will change when its y-value is increased by one unit while its z-value is held fixed.[5] So these partial regression coefficients measure the *direct* effect of each character on fitness.

In the example above, body size (z) is selectively neutral. This means that β_1 equals zero—the statistical association between z and w, reflected in the fact that Cov (w, z) is positive, disappears when we control for brain size (y).[6] So information about an organism's body size does help predict its fitness if its brain size is not already known; but if its brain size *is* known, learning its body size provides no further information about its fitness.[7] The fact that $\beta_1 = 0$ but Cov (w, z) > 0 gives formal expression to the fact that z is not subject to direct selection, but is indirectly selected as a result of its correlation with y.

Multiple regression is a standard tool for analysing situations where a variable of interest is causally affected by a number of other variables. As a technique for analysing data and drawing causal inferences, multiple regression has a number of limitations. However, our aim here is to clarify concepts, not analyse data, so not all these limitations are relevant. We can ignore limitations that have to do with estimation, that is, inferring from sample to population parameters; we assume that every organism in the population has been sampled. However, the issue of linearity requires comment.

Even if z and y are the only characters that affect fitness, it does not follow that an organism's fitness is a *linear* function of z and y, as the regression model suggests. Fitness might depend on higher-order terms such as z^2 and y^2, or on the interaction term zy. Of course, one can always *fit* the simple linear model (3.1), thanks to the residual

---

[5] The partial regression coefficients β_1 and β_2 would usually be written as $b_{w,z,y}$ and $b_{w,y,z}$, where the subscripted variable after the dot indicates the variable that is being controlled for. For ease of reading, I do not adopt this terminology here.

[6] This could alternatively be expressed by saying that the *simple* regression of fitness on z is non-zero, but the partial regression of fitness on z, controlling for y, is zero. To see that this is equivalent to the text, note that Cov (w, z) is by definition equal to b_{wz} Var (z), where b_{wz} is the simple regression of fitness on z and Var (z) is the variance of z.

[7] In philosophical terminology, the link between fitness and body size is 'screened off' by brain size. There is a close connection between the philosophical concept of screening off and the statistical concept of controlling for a variable, as Irzik and Meyer (1987) emphasize. See Chapter 4 Section 4.2 for further discussion of screening off.

term. But if the true dependence of fitness on z and y is highly non-linear, or if there is significant interaction, the residuals will be non-negligible and the partial regression coefficients β_1 and β_2 cannot be interpreted as measures of direct causal influence. Since our aim here is conceptual clarification, we assume linearity and zero interaction for simplicity; in any real case these assumptions would require empirical justification.

Now recall the simplest version of the single-level Price equation, which assumes there is no transmission bias:

$$\bar{w}\Delta\bar{z} = \text{Cov}(w, z) \qquad (3.2)$$

where $\Delta\bar{z}$ is the cross-generational change in mean character. This equation makes no reference to the character y. But we can substitute equation (3.1) into equation (3.2), to yield:

$$\bar{w}\Delta\bar{z} = \text{Cov}(\beta_1 z + \beta_2 y + e, z)$$

which simplifies to:

$$\bar{w}\Delta\bar{z} = \underset{\text{Direct selection on z}}{\beta_1 \text{Var}(z)} + \underset{\text{Indirect selection on z}}{\beta_2 \text{Cov}(z, y)} \qquad (3.3)[8]$$

Equation (3.3) is a useful way of thinking about selection on correlated characters. It expresses the total change in $\bar{z}$ as the sum of two components, the first reflecting direct selection on z itself, the second indirect selection on z as a consequence of direct selection on y. The equation is highly intuitive. That the change in $\bar{z}$ due to direct selection on z should depend on β_1 Var (z) makes good sense; for Var (z) measures the variation in z, while β_1 measures the direct effect of z on fitness. That the change in $\bar{z}$ due to indirect selection should depend on β_2 Cov (z, y) is also intuitive. For Cov (z, y) measures the correlation between z and y, while β_2 measures the direct effect of y on fitness. Therefore, if selection on y is to affect $\Delta\bar{z}$, y must both directly affect fitness and correlate with z.

In Chapter 1, we distinguished between statistical and causal decompositions of evolutionary change. What becomes of that distinction

---

[8] To see exactly how equation (3.3) is derived, first apply the distributive rule for covariance; then note that Cov (z, z) = Var (z) by definition, and Cov (z, e) = 0 by standard least-squares theory.

here? Given that we stipulated that z and y are the only characters causally affecting fitness, and that the dependence is linear, equation (3.3) can be regarded as a causal decomposition. However, in a real case we are unlikely to know that these causal stipulations are true. If so, equation (3.3) will still be correct as a statistical decomposition, but to regard it as a causal decomposition is to make a substantive hypothesis about the world. This is a strictly epistemological limitation; there must *be* a uniquely correct causal decomposition, even if any attempt to discover it is fallible.

3.3 CROSS-LEVEL BY-PRODUCTS IN MLS1

The correlated character scenario shows how a character-fitness covariance can be a by-product of direct selection on another character at the same hierarchical level. But what we are really interested in, vis-à-vis the levels of selection, is the idea that a character-fitness covariance might be a by-product of direct selection at a *different* level. This section and the next examine cross-level by-products in hierarchically organized systems.

For simplicity we assume two hierarchical levels, with strict nesting and a fixed number of particles per collective, as in Chapter 2. It may be useful to keep an abstract representation of hierarchical organization in mind; this is depicted in Figure 3.3.

Intuitively it is easy to see how cross-level by-products in the particle→collective direction might arise. For since a collective's character and fitness usually depend on, and are sometimes defined in terms of, the characters and fitnesses of its constituent particles, selection at the particle level might have effects that 'filter up' the hierarchy, generating a spurious character-fitness covariance at the collective level. The challenge is to spell out this intuitive argument in precise terms. Precisely how must character and fitness at the two levels be related

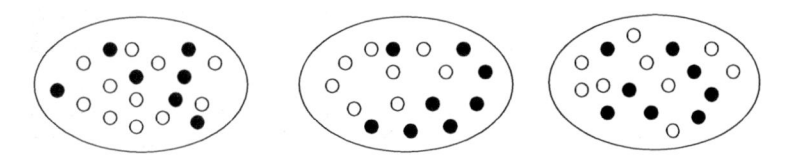

Figure 3.3. Particles nested within collectives

in order for such 'filtering up' to occur? And what about cross-level by-products running in the opposite direction? These questions needed to be addressed separately for MLS1 and MLS2, given that 'collective fitness' means something different in each; I consider them in turn.

Recall the essential features of MLS1: the particles are the 'focal' units, and collective fitness is defined as average particle fitness. Evolutionary change is judged to have occurred when the frequency of different particle-types in the *global* population of particles has changed. As we saw in Chapter 2, the evolutionary dynamics can be described by an equation in which characters and fitnesses are attributed to both particles and collectives. This was illustrated by the Price decomposition:

$$\overline{w}\Delta\overline{z} = \underset{\text{Collective-level selection}}{\text{Cov}(W, Z)} + \underset{\text{Particle-level selection}}{E(\text{Cov}_k(w, z))} \quad (3.4)$$

where Cov (W, Z) is the covariance between collective fitness and character; and $E(\text{Cov}_k(w, z))$ is the average of the within-collective covariances between particle fitness and character. Recall that for this version of Price's equation to hold true, collective character Z *must* be defined as average particle character, and collective fitness W *must* be defined as average particle fitness. Note also that equation (3.4) presumes zero transmission bias at the particle level.

An example much discussed in the group selection literature of the 1980s shows how cross-level effects can arise in MLS1 (cf. Sober 1984; Grafen 1985; Nunney 1985a; Heisler and Damuth 1987). A population of organisms is subdivided into groups. (So the collectives are groups and the particles are organisms.) The fitness of any organism depends only on its own phenotype and not which group it is in—there are no 'group effects' on organismic fitness. As Heisler and Damuth (1987) note, most biologists would say that no group selection is occurring in this situation, for the evolution of the system can be predicted *without* taking group structure into account. Nonetheless, there may well be a covariance between group fitness and group character. Some groups may be fitter than others simply because they contain a higher proportion of fit organisms.

Sober (1984) illustrated this point with an example in which an organism's fitness depends only on its own height—any two organisms of the same height have the same fitness, whatever group they are in. A group composed mainly of tall organisms, so with a high average height, will be fitter than a group composed mainly of short organisms; thus group fitness and group character covary positively. Though Sober's discussion was not framed in terms of the Price equation, his example neatly illustrates how cross-level by-products may complicate the Price approach to MLS1. In terms of equation (3.4), Sober has described a case where a non-zero value of Cov (W, Z) is a by-product of selection at the particle level.

One consequence of this is that the Price approach to MLS1 is called into question; for it detects a component of collective-level selection where intuitively there is none. This suggests that equation (3.4) may *not* in fact partition the total change into components corresponding to the two levels of selection, despite what Price and Hamilton thought. This important issue is tackled in Section 3.4.

As a way of analysing cross-level by-products in MLS1, Heisler and Damuth (1987) advocate a statistical technique from the social sciences called 'contextual analysis'.[9] The basic idea is to regard a collective's character as a 'contextual' character of each particle in the collective. So in Sober's example, average group height becomes a contextual character of every organism in the group. Each organism in the global population is therefore assigned two characters: an individual character (height), and a contextual character (average height of its group), both of which may affect its fitness. The crucial question is then: does the fitness of a particle depend only its own character, or does it also depend on its contextual, that is, collective, character?

It is important to see *why* this is the crucial question. The reason is that when a particle's fitness depends only on its own character, as in Sober's example, any character-fitness covariance at the collective level must be a by-product of selection at the particle level. (And arguably, the Price equation will mislead about the true levels of selection.) Conversely, where particle fitness *is* affected by collective character, at least some of the collective-level covariance is due to processes occurring at the collective level itself; not all of it is a side effect of lower-level selection. So determining whether particle fitness is affected by collective character provides an important clue about whether a cross-level by-product is in play.

---

[9] Boyd and Iversen (1979) provide a thorough introduction to contextual analysis, with some discussion of conceptual and philosophical issues. For applications of contextual analysis to levels-of-selection problems in biology, see Heisler and Damuth (1987), Goodnight et al. (1992), Tsuji (1995), Pederson and Tuomi (1995), Getty (1999), Okasha (2004b), and Aspi et al. (2003).

How should we determine this? Simply looking for a correlation between particle fitness and collective character is insufficient, as Heisler and Damuth stress. Even if a particle's fitness is *not* affected by its collective character, the two will still be correlated so long as particle fitness is affected by *particle* character. This is because particle character and collective character are *themselves* correlated—since the latter is defined as average particle character. (In Sober's example, tall organisms are more likely to be found in groups with high average height, obviously, so an individual's height will tend to be correlated with the average height of its group.) So we need to check whether there is a correlation between particle fitness and collective character that is *not* due to the correlation between particle fitness and particle character.

Contextual analysis addresses this question using a standard linear regression model:

$$w = \beta_1 z + \beta_2 Z + e \quad (3.5)$$

where particle fitness w is the response variable, and the two independent variables are particle character z, and collective character Z.[10] (Think of Z as a relational property of each particle in the collective.) β_1 and β_2 are the partial regression coefficients. So β_1 measures the direct effect of particle character on particle fitness, controlling for collective character; while β_2 measures the direct effect of collective character on particle fitness, controlling for particle character. For simplicity we again assume linearity, zero interaction, and the absence of unmeasured influences on particle fitness; so the partial regression coefficients can be interpreted as measures of direct causal influence.[11]

Heisler and Damuth argue that selection at the collective level requires β_2 to be non-zero. This means that information about the collective to which a particle belongs is relevant to predicting the particle's fitness, over and above information about the particle's own character, that is, it signals a 'collective effect' on particle fitness. In Sober's example, where an organism's fitness depends only on its own height, β_2 is zero—once

---

[10] Equation (3.5) could be formulated more perspicuously using indices, i.e. $w_{jk} = \beta_1 z_{jk} + \beta_2 Z_k + e_{jk}$, where w_{jk} and z_{jk} denote the fitness and character value of the jth particle in the kth collective, Z_k is the character value of the kth collective, and e_{jk} is the residual. The unindexed form is used in the text for ease of reading.

[11] The most thorough exposition of contextual analysis, due to Boyd and Iversen (1979), explicitly allows the possibility of interaction between individual and contextual characters, by adding an interaction term to the regression model. This complication, though important for any actual empirical study, will not be examined here.

you know an organism's height, further information about its average group height does not help predict its fitness.[12] Of course, if you did *not* know the organism's height, then learning the average height of its group *would* help you predict its fitness. But collective character is not a predictor of particle fitness once particle character has been taken into account. That is the key point. More generally, contextual analysis tells us that if β_2 is zero, then a non-zero value of Cov (W, Z) must be a by-product of particle-level selection.[13]

Notice that contextual analysis is simply a special case of the Lande–Arnold model of selection on correlated characters (Heisler and Damuth 1987). Indicative of this is that equation (3.5), the contextual regression model, has precisely the same form as equation (3.2). The only difference is that in the contextual model, one of the characters whose effect on particle fitness we are interested in is a 'contextual' character. As we saw, the Lande–Arnold model is meant to help tell whether a given character-fitness covariance is a by-product of selection on a correlated character. The same is true of contextual analysis. But since the two correlated characters, in the contextual model, are at different hierarchical levels, this means that the by-product in question is a *cross-level* by-product. Thanks to the expedient of treating a collective's character as a relational property of its constituent particles, contextual analysis succeeds in reducing cross-level by-products to the single-level by-products that the Lande–Arnold model analyses.

It may not be clear *how* contextual analysis achieves this reduction, for the following reason. Cross-level by-products, as defined earlier, occur where direct selection at one level produces a character-fitness covariance at another level. But contextual analysis deals exclusively with *particle* fitness—it says nothing about collective fitness. So while it is clear how contextual analysis could help determine whether a covariance between collective character and *particle* fitness is a by-product of direct selection at the particle level, this seems to be changing the concept of a cross-level by-product. The original question was whether a covariance between collective character and *collective* fitness might be a by-product of particle-level selection, but contextual analysis does not seem to address that question.

[12] The condition $\beta_2 \neq 0$ is closely related to Sober's (1984) idea that group selection requires group membership to be a 'positive causal factor' in the determination of individual fitness; see Okasha (2004c) for further discussion.

[13] *Modulo* the simplifying assumption that all of the causal influences on fitness have been captured in the contextual regression model, of course.

In fact this objection is incorrect, for the two covariances in question are identical. The covariance between collective character and collective fitness, Cov (W, Z), is equal to the covariance between collective character and particle fitness, Cov (w, Z), *given that collective fitness W is defined as the average fitness of the particles in the collective*.[14] So contextual analysis *does* address the issue of cross-level by-products as we defined them. From equation (3.5), and from the fact that Cov (W, Z) = Cov (w, Z), it follows that:

Collective-level Covariance   By-product of Selection on z   Direct Selection on Z

$$\text{Cov}(W, Z) = \beta_1 \text{Var}(Z) + \beta_2 \text{Var}(Z)$$
$$(3.6)[15]$$

Equation (3.6) is a useful decomposition for understanding by-products running in the particle→collective direction. The LHS term is the character-fitness covariance at the collective level. The first RHS term, $\beta_1 \text{Var}(Z)$, measures the cross-level by-product arising from direct selection on z. The second RHS term, $\beta_2 \text{Var}(Z)$, measures direct selection on the collective character itself. Equation (3.6) shows that the collective-level covariance is actually made up of two parts, one due to selection at the collective level itself, the other 'caused from below', by selection at the particle level.

Contextual analysis helps illuminate cross-level by-products of the sort illustrated by Sober's height example. But does it offer a complete solution to the problem of cross-level by-products in MLS1? (Contextual analysis does not apply to MLS2, since it treats the particles as the 'focal' units.) And what are its general implications for the levels-of-selection debate? These questions are tackled next.

3.3.1 Contextual Analysis: Further Remarks

Contextual analysis was introduced into the levels-of-selection discussion to deal with cross-level by-products running from lower to higher levels,

---

[14] As Heisler and Damuth (1987) say, the Cov (W, Z) term of the Price equation is 'an expectation for the individuals of the population, not for the groups themselves' (p. 585). (Read 'particle' for individual and 'collective' for group to translate this into our terminology.) This point is easily missed, given the simplifying assumption that all collectives contain the same number of particles. If that assumption were relaxed, the Cov (W, Z) term would need to be weighted by collective size, as noted in Chapter 2, which drives home the fact that it is an expectation for the particles, not the collectives.

[15] To see that equation (3.6) follows from equation (3.5), note that Cov (z, Z), the covariance between a particle's character and the character of its collective, is simply equal to Var (Z), the variance in collective character.

as in Sober's height example. However, it can equally be used to analyse by-products running in the reverse direction. For as is clear from equation (3.5), particle character z and collective character Z play symmetric roles in the regression model. Therefore, a character-fitness covariance at the particle level might be a by-product of direct selection at the collective level. (Importantly, 'direct selection at the collective level', in contextual analysis, means a direct effect of Z on w, i.e. $\beta_2 \neq 0$.) This would happen if the fitness of a particle depends only its collective character, and not at all on its own character; so any two particles within the same collective have identical fitness. In terms of equation (3.5), this means that $\beta_1 = 0$ but $\beta_2 \neq 0$.

More generally, we can use contextual analysis to express the overall particle-level covariance as the sum of two components, just as we did for the collective-level covariance. This yields:

Particle-level covariance   Direct selection on z   By-product of selection on Z

$$\text{Cov}(w, z) \quad = \quad \beta_1 \text{Var}(z) \quad + \quad \beta_2 \text{Var}(Z)$$
(3.7)

which is a useful decomposition for thinking about by-products in the collective→particle direction. The LHS term is the overall covariance between particle fitness and particle character, in the global population of particles. The first term on the RHS, $\beta_1 \text{Var}(z)$, reflects direct selection on z itself, while the second term, $\beta_2 \text{Var}(Z)$, is the cross-level by-product arising from direct selection on Z. If a particle's fitness is unaffected by its own character, that is, if $\beta_1 = 0$, then the entirety of the particle-level covariance must be a side effect of selection at the collective level. Conceptually this is exactly the reverse of Sober's height example: the cross-level by-product runs from higher to lower.

Causal graphs permit a useful representation of cross-level by-products running in both directions. In Figure 3.4, diagram (A) depicts a particle→collective by-product, as in Sober's example. Particle fitness w is causally affected by particle character z but not by collective character Z; since z and Z are themselves correlated, this results in a spurious covariance between w and Z. Diagram (B) depicts the reverse: w is causally affected by Z but not by z, resulting in a spurious covariance between w and z.[16]

---

[16] Note that these causal graphs are isomorphic to the graph used to depict selection on correlated characters in Figure 3.2, which reflects the point that contextual analysis is a special case of the Lande–Arnold model.

Causality and Multi-Level Selection

```
(collective character)  (particle fitness)        (collective character)  (particle fitness)
      Z------------w                                     Z------------→w
       \          /\                                      \          /
        \        /                                         \        /
         \      /                                           \      /
          \    /                                             \    /
           \  /                                               \  /
        z (particle character)                          z (particle character)

  (A) Particle→collective by-product              (B) Collective→particle by-product
```

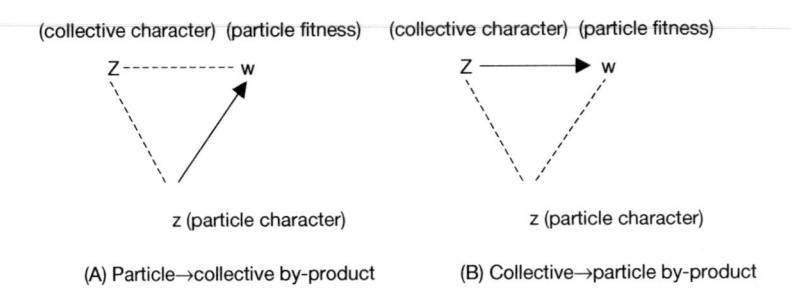

Figure 3.4. Cross-level by-products in MLS1

Note that in these graphs collective fitness W does not feature explicitly; the graphs depict causal influences on particle fitness. But since W is defined as average particle fitness, the graphs *do* convey information about the character-fitness covariance at the collective level. For recall that Cov (W, Z) = Cov (w, z), given the definition of collective fitness. So in diagram (A), there *is* a cross-level by-product as defined earlier, that is, a character-fitness covariance at one level resulting from direct selection at another. Things are slightly more complicated with diagram (B). There is certainly a by-product, in that Cov (w, z) assumes a non-zero value as a side effect of the causal influence of Z on w. But it might be argued that this is not a *cross-level* by-product, since a causal influence of Z on w is not what 'direct selection at the collective level' is supposed to mean. Surely 'direct selection at the collective level' should mean a direct effect of collective character on *collective* fitness, not particle fitness?

This objection is not a good one, presuming we agree that MLS1 does constitute real multi-level selection. For in MLS1, collective fitness is a logical construct out of particle fitness. So any causal factors that affect the fitness of a collective can *only* do so by affecting the fitnesses of the particles it contains; there is no other way to affect a collective's fitness. Put differently, given a complete model of the factors affecting particle fitness, including collective characters, whether there is a non-zero covariance between collective fitness and collective character *is determined as a matter of logic*.[17] So if we are to avoid the conclusion

---

[17] This can be seen from equation (3.7), which tells us that the value of Cov (W, Z) is fully determined by β_1, β_2, and Var (Z). The point to note is that β_1 and β_2 measure direct effects on *particle* fitness, while Var (Z) is simply the variance of the collective character.

that character-fitness covariance at the collective level is *always* a by-product of selection at the particle level, a conclusion which would make multi-level selection of the MLS1 variety impossible, the only solution is to think of collective-level selection as selection on the component of particle fitness that is determined by collective character, as does contextual analysis.

The fact that contextual analysis treats by-products running in both directions symmetrically may seem surprising, given that the properties of wholes normally depend on the properties of their parts, rather than vice versa. But it is easily explained given what contextual analysis means by collective-level selection. That a particle's fitness may be causally affected by properties of its host collective involves no metaphysical mystery; it is simply an instance of the familiar fact that the fitness of any biological unit is environment-relative. So while contextual analysis involves a sort of downward causation, it is not a metaphysically onerous sort.

In our exposition of contextual analysis, we considered just one particle character, z, and one collective character, Z, defined as average particle character. This was strictly for ease of exposition. In a real multi-level scenario, it is likely that particle fitness will be affected by many particle characters and many collective characters. Extending the contextual model to include multiple characters is conceptually straightforward. Heisler and Damuth (1987) make the important point that emergent as well as aggregate collective characters can be included in the contextual regression model. A particle's fitness could easily be affected by emergent characters of its collective, for example, an ant's fitness might depend on the amount of division of labour within its colony, which is an emergent colony-level trait. Note, however, that the character z necessarily correlates with average particle character Z, but not necessarily[18] with any other collective characters. So direct selection on z will always lead Z to correlate spuriously with fitness; the same is not true for other collective characters.[19]

Since contextual analysis treats cross-level by-products running in both directions in symmetric fashion, and says nothing about which are more likely to be found in nature, it may seem neutral with respect to

---

[18] Except in the limiting case where all the collectives have the same Z-value.

[19] Note that equation (3.6) above, which expresses Cov (W, Z) as the sum of two components, assumes that the collective character Z is defined as average particle character. The corresponding equation for emergent collective characters is Cov (W, Z) = β_1 Var (Z) + β_2 Cov (z, Z); this equation holds true however Z is defined.

the reductionism–holism issue. In a sense this is right. But in another sense contextual analysis is intrinsically reductionist, simply because it deals with MLS1 rather than MLS2. Any approach to selection in hierarchical systems that takes the particles to be the 'focal' units, and treats the collectives as part of their fitness-affecting environments, is thereby committed to a 'bottom up' mode of explanation. Boyd and Iversen (1979) say that the aim of contextual analysis in the social sciences is 'to understand complex societal-level phenomena in terms of the motives and behaviours of individuals', an aim which is explicitly anti-holistic (p. 25); the same is true of contextual analysis in multi-level selection theory.

3.4 CONTEXTUAL ANALYSIS VERSUS PRICE'S EQUATION

Importantly, contextual analysis constitutes a rival to the Price equation approach to MLS1, outlined in Chapter 2. This section explores the opposition between the two approaches.

The essence of the Price approach is enshrined in equation (3.4) above, which Hamilton (1975) described as effecting a 'formal separation of levels of selection' (p. 333). But the example used to motivate contextual analysis originally, where a particle's fitness depends only on its own character, calls Hamilton's claim into question. For in that example, the Price approach detects selection at the collective level where intuitively there is none. One might conclude from this that the Price approach is theoretically flawed, as do Nunney (1985a), and Heisler and Damuth (1987). This is quite a radical conclusion, since the Price approach *seems* to capture our pre-theoretic understanding of selection at multiple levels very neatly, and is widely used among multi-level selection theorists.

The tension between contextual analysis and the Price approach arises because they constitute non-equivalent ways of partitioning the total evolutionary change into components corresponding to each level of selection. For ease of comparison, assume for the moment that the collective character Z *is* defined as average particle character. The respective partitions are then:

		Collective-level selection	Particle-level selection
Price:	$\overline{w}\Delta\overline{z} =$	$\text{Cov}(W, Z)\ +$	$E(\text{Cov}_k(w, z))$
Contextual:	$\overline{w}\Delta\overline{z} =$	$\beta_2 \text{Var}(Z)\ +$	$\beta_1 \text{Var}(z)$

Note that the Price partition is simply equation (3.4); the contextual partition is derived by substituting equation (3.7) above into the simple Price equation for the particles. Again, zero transmission bias at the particle level is assumed.

The key difference between the two approaches can be seen by inspecting the term for collective-level selection. On the Price approach, this term is Cov (W, Z), the collective-level character-fitness covariance. But as we have seen, contextual analysis holds that this covariance is made up of two parts, one of which is a cross-level by-product and must thus be subtracted to yield a true separation of the levels of selection (cf. equation (3.6)). Once this subtraction is made, the amount left over is β_2 Var (Z), which is what contextual analysis attributes to collective-level selection.[20] So we can regard the contextual partition as a 'correction' of the Price partition that results from taking explicit account of cross-level by-products.

The Price and contextual partitions are both correct as *statistical* decompositions of the total change, for both of the above equations are true; but at most one of them can constitute a correct *causal* decomposition. In other words, presuming there is a 'fact of the matter' about how much of the total change is attributable to selection at each level, at most one of the equations captures that fact. For the two equations will always divide up the total change differently;[21] and in certain cases they will disagree about whether there is *any* component of selection at one of the levels. So they embody conflicting conceptions of multi-level selection.

How should we choose between the Price and contextual approaches? This is a tricky question. Contextual analysis seems superior on theoretical grounds. If we think of particle selection and collective selection as separate evolutionary 'forces', each capable of influencing the evolution of a trait, as biologists often do, then the use of partial regression techniques is surely appropriate. Moreover, in the case where a particle's fitness depends only on its own character, contextual analysis generates the intuitively correct answer—that all the selection is at the particle

---

[20] Note that this does *not* imply that contextual analysis will always attribute a smaller component of the total change to collective-level selection than the Price approach, for the subtracted component may be positive or negative.

[21] Except in the degenerate case where all the groups have the same Z-value. In all non-degenerate cases, Cov (W, Z) and β_2 Var (Z) must differ in value, as is clear from equation (3.6), hence the two approaches attribute different fractions of the total change to collective-level selection.

level—while the Price approach does not. These are both points in favour of the contextual approach.

However, the contextual approach has one implication that some theorists find deeply counter-intuitive. As Heisler and Damuth (1987) and Goodnight et al. (1992) observe, in the scenario known as 'soft selection', contextual analysis detects a component of selection at the collective level, which is intuitively wrong. Soft selection means that all the collectives have identical fitness, that is, contribute the same number of particles to the next generation (Figure 3.5). (This could occur if each collective's output is limited by resource constraints.) However, there are 'collective effects' on particle fitness. The fitness of a particle depends not just on its own character, but also on its relative character ranking within its collective. Thus suppose the trait undergoing selection is size. Large particles have a fitness advantage relative to small ones within the same collective, that is, they gain a larger share of the collective's reproductive output. So any particle, large or small, benefits from being in a collective of low average size. Particle fitness thus depends on both particle and collective character.

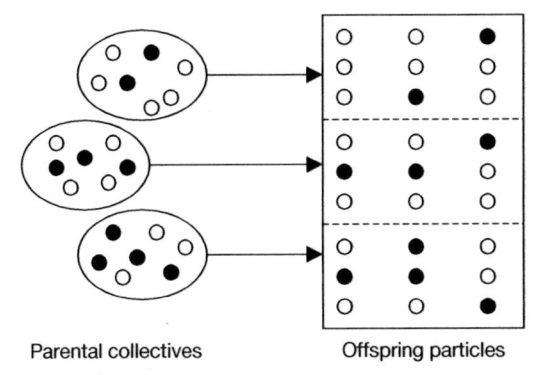

Parental collectives   Offspring particles

Note that all the parental collectives contribute equal numbers of offspring particles

Figure 3.5. Soft selection

The soft selection example elicits different verdicts from the Price and contextual approaches. Since all the collectives have the same fitness, it follows immediately that Cov (W, Z) is zero—if collective fitness doesn't vary, it cannot *co*vary with anything—so the Price approach says there is no collective-level selection. However, the contextual approach detects a component of selection at both levels. For since particle fitness depends on both particle and collective character, both β_1 and β_2 in the contextual regression model are non-zero; so both components of the contextual partition will be non-zero.[22] Therefore, contextual analysis says there is selection at the collective level *even though* all the collectives have the same fitness.

Intuitively this is a strange result. For the idea that selection at a level requires variance in fitness at that level is virtually axiomatic. This idea is an integral part of the Lewontin conditions, which form the starting point for almost all discussions of the levels of selection, and is enshrined in the Price approach. But the contextual approach violates this idea. In the soft selection case, the fact that particle fitness is affected by collective character is sufficient for there to be a component of selection at the collective level, according to contextual analysis, even without any variance in collective fitness. So in this case it is the Price approach, not the contextual approach, that generates the intuitively correct result.

Since contextual analysis implies a violation of the Lewontin conditions, one might conclude that it has distorted the intuitive concept of multi-level selection beyond recognition. D.S. Wilson (personal communication) argues just this; see also Rice (2004). But this conclusion does not necessarily follow. For arguably, the concept of multi-level selection that the Lewontin conditions capture, at least in the first instance, is MLS2—where collective fitness is understood in terms of collectives 'making more' collectives. But the tension between the contextual and Price approaches arises within the MLS1 framework, where the particles are the focal units. So accepting the contextual approach to MLS1 would not mean abandoning the revered Lewontin conditions altogether; it would mean restricting those conditions to multi-level selection of the MLS2 variety only, which is anyway their natural home (Okasha 2004b).

This consideration mitigates the counter-intuitiveness of what contextual analysis says about soft selection, but does not provide a fully satisfactory resolution. For a number of theorists whose concern is MLS1, and who are clearly aware of the distinction between MLS1

---

[22] Presuming that the collectives do not all have the same character value, i.e. Var (Z) $\neq 0$. Note that in soft selection, β_1 must equal minus β_2, i.e. the partial regression of fitness on particle character must be equal but opposite in sign to the partial regression of fitness on collective character (Goodnight et al. 1992). This follows directly from equation (3.6) above, given that part of the definition of soft selection is that Cov (W, Z) = 0.

and MLS2, insist that collective-level selection requires variance in collective fitness. For example, Sober and Wilson (1998), in discussing how altruism can spread by group selection of the MLS1 type, emphasize repeatedly that group selection requires variance in group fitness. Similarly, Wade, Goodnight, and Stevens (1999), in a discussion of interdemic selection, acknowledge that 'most would consider [variance in demic productivity] a prerequisite for interdemic selection'; but as they note, contextual analysis implies that it is not a prerequisite (p. 600). So appealing to the distinction between MLS1 and MLS2 does not defuse the unpalatable result of the contextual approach entirely.

The tension between the Price and contextual approaches arises because there are two requirements that, pre-theoretically, we would like satisfied in order for there to be selection at the collective level. The first requirement is a 'collective effect' on particle fitness. As Sober's height example shows, where this requirement is not satisfied, particle-level selection seems to be doing all the causal work. The second requirement is variance in collective fitness. As the soft selection example shows, where this requirement is not satisfied we are reluctant to acknowledge collective-level selection, since the Lewontin conditions are not satisfied. The Price and contextual approaches each satisfy one of these requirements but not the other.

In his well-known treatment of group selection, Sober (1984) tries to integrate both requirements. He simply *defines* group selection as a causal process which occurs when certain conditions are satisfied; the conditions include variance in group fitness and a 'group effect' on individual fitness (p. 280). This move may seem attractive, but is ultimately unsuccessful. For Sober's definition only addresses the *qualitative* question 'is group selection occuring, or not?'; it leaves untouched the *quantitative* question 'how *much* of the total evolutionary change is due to group selection?', which is the more fundamental issue. In a case where Sober's definition says that group selection *is* occurring, we might still be interested in the quantitative question—to which the Price and contextual approaches give conflicting answers. Sober's definition, though well-motivated, does not resolve the conflict.

A quite different way of trying to reconcile the two requirements is to impose constraints on what counts as a collective. Neither of the two partitioning techniques—Price or contextual—involves a commitment to any particular definition of a collective; each would work perfectly well even if the particles were divided into collectives in a purely arbitrary way. In Chapter 2, we discussed the idea that collectives should be defined on the basis of fitness-affecting interactions among the particles; it is these interactions that confer 'biological reality' on the collectives, according to many theorists.

This 'interactionist' conception of a collective, *if combined with the Price approach*, offers a way of reconciling the two requirements on collective-level selection. For recall the case where a particle's fitness depends only on its own character—the problem case for the Price approach. Given the interactionist conception, it follows that no collectives actually exist in such a case—the global population of particles is hierarchically unstructured. For if the fitness of each particle depends only on its own character, then obviously there are no fitness-affecting interactions between the particles. The problem with the Price approach is thus solved *without* using contextual analysis, by taking care to specify what counts as a 'collective'. The resulting conception of MLS1 satisfies both pre-theoretic requirements. Collective-level selection requires both variance in collective fitness, for otherwise Cov (W, Z) is zero, and 'collective effects' on particle fitness, for otherwise there are no collectives in the first place. So the Price approach can be salvaged after all.

This solution is attractive, but faces two problems. First, the interactionist conception of a collective is not universally applicable, as noted in Chapter 2. Secondly and more importantly, the theoretical argument in favour of contextual analysis is not *just* that it yields the 'correct' answer in the limiting case where a particle's fitness depends only on its own character. The failure of the Price approach to get the answer right in this case is symptomatic of a deeper problem, namely that the term Cov (W, Z) will *always* contain an element which is a by-product of particle-level selection. This remains true however collectives are defined. So although employing the Price approach in conjunction with the interactionist conception avoids the problem in the limiting case, the theoretical objection to the Price approach remains.

Finally, consider a conventionalist dissolution of the problem. Perhaps there is 'no fact of the matter' about whether the Price or the contextual approach is correct? Since both yield correct statistical decompositions of the total change, perhaps it is a mistake to ask which captures the 'real' causal facts? The problem with this suggestion is that the distinction between statistical and causal decomposition is not a philosophical artefact; it is implicit in the way biologists actually speak. Biologists frequently ask which level of selection in a multi-level system is likely to predominate. This presumes that it *is* possible to partition the total change into components corresponding to the different levels of selection—for otherwise it makes no sense to talk about the relative magnitude of selection at each level. Of course, the fact that biologists make a certain presumption does not mean it is true. But it does make the conventionalist dissolution an option of last resort.

In my view, the contextual approach is on balance preferable, despite the violation of the Lewontin conditions that it entails; for the Price approach cannot deal satisfactorily with cross-level by-products. If this is correct then a certain amount of conceptual revision is called for, for the Price approach underpins the way many theorists think about multi-level selection, and is heuristically valuable. But the issue is not clear-cut. For as we shall see when we discuss 'genic selection' in Chapter 5, there are multi-level scenarios of the MLS1 variety which the contextual approach cannot satisfactorily handle, but the Price approach can.

Two further differences between the Price and contextual approaches to MLS1 deserve mention. First, on the contextual approach, collective-level selection can operate on *any* collective character, aggregate or emergent, that causally affects fitness. But on the Price approach, the collective character Z *must* be defined as average particle character, on pain of the partitioning technique not working. This is a significant limitation, given that there is no empirical reason why average particle character should be accorded a privileged role over any other collective character in the selection process. In this respect, the contextual approach provides a representation of multi-level selection that is inherently more general.

Secondly, the Price approach requires that the particles be strictly nested in collectives, that is, with no overlapping. The contextual approach, as expounded above, also assumes strict nesting, but unlike the Price approach it can easily be modified to deal with overlapping. With overlapping, some particles are members of more than one collective, so have two or more sets of collective characters, rather than just one set; this simply lengthens the list of collective characters that can potentially affect particle fitness, so is easily handled by the contextual regression model. In Chapter 2, we left open the question whether strict nesting is a well-motivated requirement on part–whole structure. If the argument of this section—that the contextual approach to MLS1 is superior to the Price approach—is correct, this suggests that strict nesting is not well-motivated. For the causal process that contextual analysis defines as collective-level selection—differences in particle fitness that arise from differences in collective character—may occur whether or not the particles are strictly nested in collectives.

3.5 CROSS-LEVEL BY-PRODUCTS IN MLS2

Recall the essential features of MLS2: particles and collectives are both 'focal' units, and their fitnesses are independently defined, usually over different timescales. Selection at the particle level affects the frequency of different particle types within collectives; selection at the collective level affects the frequency of different collective types. This requires that the collectives engage in bona fide reproduction, that is, that they 'make more' collectives.

3.5.1 Particle→Collective By-Products

Cross-level by-products in the upward direction are easily envisaged in MLS2. For within each collective, selection occurs among the particles. This could affect any collective character, aggregate or emergent, which depends in a suitable way on the characters of the constituent particles. One emergent character of a collective is its fitness$_2$, that is, its number of offspring collectives. So in principle, a character-fitness covariance at the collective level, of the MLS2 sort, could be a side effect of lower-level selection. The challenge is to make this intuitive argument precise.

Importantly, particle→collective by-products in MLS2 are fundamentally different from their counterparts in MLS1, though they admit of a similar informal characterization. The key difference is that in MLS2, collective fitness Y is *not* defined as average particle fitness, so the character-fitness covariance at the collective level, Cov (Y, Z), is *not* equal to Cov (w, Z), as it is in MLS1.[23] This means that cross-level by-products in MLS2 are *not* a special case of the single-level by-products that arise from selection on correlated characters, as in MLS1.

To fix ideas, consider some putative examples of cross-level by-products in MLS2. In her critique of species selection, Vrba (1989) argues that genuine species selection requires 'causation at the focal level of species', a requirement she thinks is rarely satisfied. More often, differential rates of extinction/speciation are side effects of 'Darwinian selection of organisms and other lower level processes' (p. 130–1). Vrba illustrates this point with an example of two African antelope clades, one containing ecological specialists (stenotopes), the other ecological generalists (eurytopes). Species in the former clade were fitter, that is,

---

[23] Indeed, these two covariances are not even in the same units.

left more offspring species, so there was a character-fitness covariance at the species level. But the explanation for this, Vrba argues, is that organismic selection produces more local differentiation in stenotopes than in eurytopes, which in turn promotes speciation. So the differential speciation was a by-product of organismic selection, not the result of a causal process at the species level. Similarly, Maynard Smith (1983) argues that the competitive replacement of one group of species by another is usually not true species selection, since 'the replacement occurs because individuals of the new taxon were competitively superior to individuals of the old one' (p. 137); Sterelny (1996a) argues likewise.

Another example comes from Michod's (1999) work on the evolution of multicellularity. Michod argues that two levels of selection were involved in the evolution of the first multicelled organisms. Selection occurred between cells within organisms, owing to variant cell-types dividing at different rates, and between the organisms themselves, owing to differences in the number of offspring they left. (Organisms are assumed to reproduce by sending out propagules containing one or more cells, which then develop and grow by cell division; see Figure 3.6.) Michod's model of this process is of the MLS2 sort: organism and cell fitness are defined in different units, and across different timescales. A cell's fitness is its rate of division within its host organism; an organism's fitness is its rate of propagule production. However, Michod and Roze (1999) argue that for organisms 'on the threshold of multi-cellular life', all differences in organism fitness may be *due* to differences in cell fitness (p. 10). An organism composed of very fit cells, that divide fast, will achieve a large adult size; and adult size may determine the number of offspring propagules that it leaves. For Michod and Roze, this means that the organism lacks true 'individuality'; in our terms, it is a cross-level by-product from the cell to the organism level.

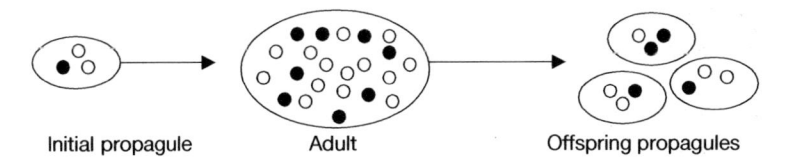

Figure 3.6. Life cycle in Michod's model for the evolution of multicellularity

How should we characterize particle→collective by-products in MLS2 in general terms? Suppose that Cov (Y, Z) is non-zero, that is, collective fitness$_2$ covaries with a given collective character Z. The issue is whether this covariance reflects causality at the collective level, or is 'caused from below'. One natural suggestion is this. If a collective's fitness$_2$ is causally determined by the average fitness of its constituent particles W, (i.e. its fitness$_1$), then there is a cross-level by-product in play. For this means that Z is not causally influencing Y—the statistical association between Z and Y disappears when we control for W. Since W is average particle fitness, lower-level fitness differences are doing all the causal work.

If this suggestion is right, then particle→collective by-products in MLS2 can be represented by the causal graph shown in Figure 3.7. As before, the thick arrow indicates causal influence, the dotted lines correlation. Note that the causal graph says nothing about *why* Z and W are correlated. But one possible reason is that within each collective, some particle character z, on which the collective character Z depends, directly affects particle fitness w; as a result, collectives with a high W-value also have a high Z-value. Fleshing out the graph this way gives substance to the idea that particle-level selection is doing all the causal work. But the key idea is contained in the graph itself: where Y is wholly determined by W, a cross-level by-product is in play. So genuine collective-level selection of the MLS2 sort, that is not a by-product of particle-level selection, requires that not all differences in collective fitness$_2$ stem from differences in fitness$_1$; they must stem in part from differences in the character Z.

(collective character)   (collective fitness$_2$)

Z - - - - - - - - - - - - - - - - - - - ▶ Y

W (collective fitness$_1$, i.e. average particle fitness)

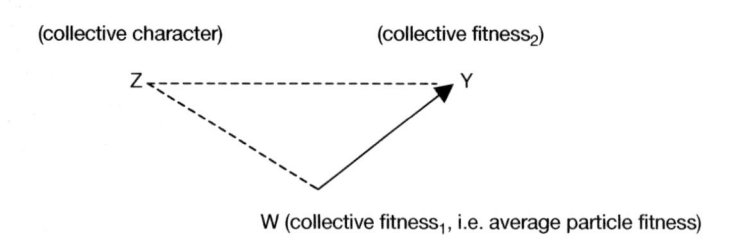

Figure 3.7. Particle→collective by-products in MLS2

As before, we can write a linear regression equation that corresponds to the causal graph:

$$Y = \beta_1 Z + \beta_2 W + e$$

where β_1 and β_1 are the partial regression coefficients, which, modulo the assumptions noted previously, may be construed as measures of direct causal influence. We can then partition Cov (Y, Z) into two components:

Collective-level covariance		Direct selection on Z		By-product of selection on W
Cov (Y, Z)	=	β_1Var (Z)	+	β_2Cov (W, Z)

(3.8)

The first term on the RHS, $\beta_1\mathrm{Var}(Z)$, reflects direct selection on the collective character Z. The second RHS term, $\beta_2\mathrm{Cov}(W, Z)$, is the by-product resulting from selection on W; according to the suggestion we are considering, this is a *cross-level* by-product, reflecting the effects of particle-level selection filtering up the hierarchy.

How plausible is this suggestion? It seems to capture at least some of the examples from the literature. The case discussed by Maynard Smith and Sterelny, of the competitive replacement of one taxon by another, has essentially the above structure. If one species drives another extinct because it contains fitter individuals, then species fitness is likely to be causally determined by average organismic fitness. If we assume that members of the first species are fitter because they are taller, then the species character 'average height' will covary with species fitness. But the covariance will disappear once we control for average organismic fitness. Similarly in Michod and Roze's example. If some organisms leave more propagules than others but only because they contain faster dividing cells and are thus larger, this means that organismic fitness is causally determined by average cell fitness. So a character-fitness covariance at the organism level, for example, between organismic fitness and the character 'number of fast dividing cells in the propagule', will be a side effect of selection at the cellular level—it will disappear once we control for average cell fitness.

Despite neatly fitting these examples, our suggested characterization of particle→collective by-products in MLS2 cannot be the whole story. Seeing whether the collective-level character-fitness covariance disappears when we control for average particle fitness provides a useful clue as to whether particle-level selection is doing the causal work, but not a definite answer. For the effects of particle-level selection, acting within a collective, will not *always* manifest themselves in the average particle fitness; so even if particle-level selection *is* causally responsible, controlling for average particle fitness may not detect this. So our characterization provides only a sufficient condition for a particle→collective by-product, not a necessary condition. If differences in collective fitness$_2$ are entirely due to difference in fitness$_1$, then all the selection is at the particle level, but not conversely.

To appreciate why the condition cannot be necessary, note that natural selection (at a single level) does not always increase the mean fitness of the population. (It is sometimes thought that Fisher's (1930) 'fundamental theorem' asserts otherwise, but this is now known to be a misinterpretation. His theorem was *not* about the change in mean population fitness.[24]) Especially where fitnesses are frequency-dependent, selection can drive the mean population fitness down, or leave it unchanged. So in a multi-level scenario, if particle-level selection is occurring within a collective, this need not manifest itself by affecting the collective's fitness$_1$; so asking whether differences in collective fitness$_1$ fully explain differences in fitness$_2$ is not a sure-fire test for whether particle-level selection is doing all the causal work. For it is possible that particle-level selection *is* doing the work, but not manifesting itself through differences in collective fitness$_1$.

What would a full account of particle→collective by-products in MLS2 look like? This is a difficult question to which I do not have an answer. For the moment, I want to look at one possible answer which, though flawed, is highly revealing. In discussing species selection, Vrba (1989) and Eldredge (1989) both flirt with an argument which, if accepted, would make genuine collective-level selection in an MLS2 scenario extremely rare. For taken to its logical conclusion, the argument implies that *any* collective-level character-fitness covariance can ultimately be 'explained from below', so an autonomous selection process at the collective level is impossible. (Neither Vrba nor Eldredge endorses this conclusion outright, so what follows should not be taken as an exposition of their views.)

---

[24] Fisher was talking about the *partial* change in mean population fitness that arises from changes in allele frequency, when the 'average effects' of the alleles are assumed constant (cf. Ewens 1989, Frank 1995a, Edwards 1994). An allele's 'average effect' measures its effect on genotype fitness; it equals the partial regression of genotype fitness on allelic 'dosage'. Fisher described this partial change (which must be non-negative) as the change 'due to' selection, but as Ewens (1989) notes, this interpretation is dubious. For the average effects usually depend on allele frequencies, and selection changes these frequencies; so the change 'due to' natural selection can hardly be identified with the partial change that results from holding the average effects constant. See Grafen (2003) for a defence of Fisher's interpretation.

The argument goes like this. In general, properties of collectives are likely to depend systematically on the properties of their constituent particles; this is related to what metaphysicians call 'part–whole supervenience', and was discussed in Chapter 2. So any collective character Z will be 'realized' by some complex of underlying particle characters; the same is true of collective fitness Y. Therefore, there cannot be a direct causal influence of Z on Y. Any apparent causal link between Z and Y is actually a side effect of causal connections between the respective particle-level characters that realize Z and Y. So genuine collective-level selection, that is irreducible to causal processes acting at the particle level, is impossible; it flies in the face of the fact that properties of collectives are determined by properties of their constituent particles.

This argument is depicted graphically in Figure 3.8. The solid arrows and dotted lines denote causation and correlation respectively; the thick vertical lines denote the relation of determination, or supervenience. So the diagram suggests that all the causality is at the particle level; the correlation between Z and Y is spurious. Let us call this the 'supervenience argument' against the possibility of genuine collective-level selection.

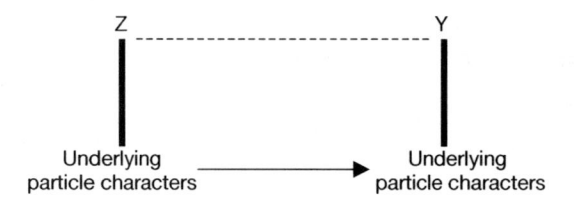

Figure 3.8. The supervenience argument against collective-level selection

What should we make of the supervenience argument? Clearly, it threatens to make particle→collective by-products ubiquitous, for it challenges the very idea of higher-level causation in a hierarchical system. However, note that the supervenience argument, if correct, shows only that a character-fitness covariance at the higher level must be a by-product of *some lower-level causal processes or other*, not necessarily lower-level *selection*. For the underlying particle characters on which Y supervenes will not necessarily be particle fitnesses; they may be characters of any sort. By contrast, our attempted characterization of particle→collective by-products, above, explicitly looked at the relation between Y and particle fitness.

I think the concept of a cross-level by-product should be restricted to a character-fitness covariance at one level that results from *selection* at another level, rather than just some causal process or other, for the former is what matters in the context of the levels of selection. After all, on plausible metaphysical assumptions we can always expect a 'micro-causal' explanation of any given character-fitness covariance. For example, if eating vitamins improves human longevity, hence fitness, there will presumably be some physiological explanation of why this is. The supervenience argument says that the higher-level covariance is spurious, for it must have a complete micro-causal explanation. But the micro-casual explanation will not necessarily involve *selection* at the lower level, which is the interesting possibility.[25]

This conclusion is reinforced by observing that the supervenience argument depends only on the fact that collectives' properties depend on the properties of their parts; it makes no difference whether the 'parts' are candidate units of selection at all. Indeed, a similar argument has been much discussed in philosophy of mind, where natural selection is not at issue. Kim (1998) argues that since mental properties (e.g. 'being angry') supervene on physical properties (e.g. 'being in such-and-such a neurological state'), it is not possible for the former to be causally efficacious; the real causal responsibility is always borne by the physical properties. Few philosophers find this conclusion palatable, but it is not easy to say where Kim's argument goes wrong.[26] The issue cannot be tackled here; the point to note is just that the supervenience argument is identical in all relevant respects to Kim's. The argument's validity thus turns on general issues in metaphysics, not ones specific to the levels-of-selection question.

However, one further parallel with the philosophy of mind deserves mention. A recurrent idea in the literature is that some mental properties are 'emergent', so in a sense irreducible to a physical basis; it is sometimes thought that this provides an escape route from Kim's argument.[27] Interestingly, evolutionists impressed by the supervenience argument, but aware that it threatens to make real collective-level selection impossible, have made an analogous move. Thus Vrba (1989)

---

[25] Thus Vrba is wrong to talk about 'Darwinian selection of organisms *and other lower level processes*' in her description of organism→species by-products; this runs together two different phenomena (1989 p. 131, my emphasis).

[26] See Loewer (2002) for a good discussion of the options.

[27] See the papers collected in Beckerman, Flohr, and Kim (eds.) (1992) for discussion of this move.

allows that a causal influence of species character on species fitness *may* be possible if the character in question is emergent; Nunney (1993) defends a similar idea. I suggest that this is a mistake. If the question at stake is whether particle-level *selection* bears the causal responsibility for a character-fitness covariance at the collective level, then emergence is irrelevant. If the question is whether *some particle-level causal processes or other* bear the causal responsibility, then emergence *may* be relevant; this depends on broader metaphysical issues. However, the first of these questions is the important one, not the second. It seems likely that the two questions have been conflated by those who favour the emergent character requirement on higher-level selection. Emergence is discussed further in Chapter 4.

Where does this leave us? An adequate account of particle→collective by-products in MLS2 must satisfy two desiderata. First, it must avoid the supervenience argument; it must only diagnose a by-product when particle-level *selection* is doing the causal work. Secondly, it must respect the fact that particle-level selection can manifest itself in a myriad of ways, not solely by affecting average particle fitness. How to provide such an account I do not know, but some progress has been made—we have identified a sufficient condition for the existence of by-products of this type.

3.5.2 Collective→Particle By-Products

Finally, what about cross-level by-products in the downward direction? Could selection at the collective level, of the MLS2 sort, produce a character-fitness covariance at the particle level? Some care is needed with this idea. Recall that in MLS2, particle-level selection takes place *within* each collective, so there is a different particle-level character-fitness covariance for each collective. As before, we let $Cov_k(w, z)$ be the covariance between particle character z and particle fitness w within the kth collective. If we like, we can also regard the particles as comprising a global population, without regard to collective membership, and calculate a single character-fitness covariance for this global population. As before, we let $Cov(w_i, z_i)$ denote this covariance. In speaking of 'character-fitness covariance at the particle level' we must be careful to specify whether we mean the within-collective or the global covariance.

Consider the within-collective covariance first. Could a non-zero value of $Cov_k(w, z)$ be a by-product of selection at the collective level? This is hard to envisage. Suppose for simplicity that the selection is by differential survival—some collectives survive while others die. How could such a mechanism affect the relationship between w and z *within* any of the surviving collectives? For since Cov_k (w, z) refers solely to the kth collective, it is hard to see how its value could be affected by a causal process involving *other* collectives.

This is correct, but not the whole story. For note that Cov_k (w, z) can be regarded as a complex character of the kth collective itself, logically on a par with its other characters. In general, if we are trying to explain why an entity has a given character, we cite the causal factors that led the entity to acquire that character. Such 'developmental' explanations are usually non-selective. As Sober (1984) notes, natural selection explains the prevalence of a trait in a population, not its presence in individual entities. So it is true that collective-level selection cannot explain why any one collective has the particular value of Cov_k (w, z) that it does—this calls for a developmental explanation. However, collective-level selection might explain why some values of Cov_k (w, z) are more prevalent than others in the population of collectives.

In effect, this is to suggest that the character Cov_k (w, z) might *itself* be subject to selection at the collective level, that is, might causally influence collective fitness Y. This idea is by no means far-fetched. Cov_k (w, z) refers to the covariance between z and w among the particles within the kth collective, so it measures the amount of selection that will occur within that collective. This might well affect the collective's fitness. Buss (1987) and Michod (1999) both argue that the evolution of cohesive multicelled organisms required mechanisms for suppressing within-organism selection; the same may be true of eusocial insect colonies and even human societies (Frank 1995b, Heinrich et al. 2003). In effect, this is to say that collectives with a value of Cov_k (w, z) close to zero enjoy a fitness advantage, for they do not get disrupted by selection among the particles contained within them. So selection at the higher level exerts a control on the amount of lower-level change possible.

If collective-level selection does act on the character Cov_k (w, z), should this be classified as a cross-level by-product? In my view it should not. The essence of a cross-level by-product is 'causal trumping'—a character-fitness covariance at one level *appears* to be causal but turns out to be spurious, trumped by a selection process at a different level. But if collective-level selection acts on the character Cov_k (w, z), favouring some values of that character over others, this will *not* generate a spurious covariance. Pick a surviving collective from the post-selection population and consider its value of Cov_k (w, z). It is quite possible that this covariance *does* reflect a direct causal connection between z and w. The fact that the collective has survived the selective filter, and has done so in virtue of the value of Cov_k (w, z) that is has, in no way implies that this covariance is spurious. So while the possibility that collective-level selection might act on the within-collective character-fitness covariance is theoretically interesting, it does not constitute a cross-level by-product. Rather, it is a type of *interaction* between levels of selection.

What happens if we consider the character-fitness covariance in the global population of particles, that is, Cov (w_i, z_i)? Might *this* covariance be a by-product of selection on the collectives, of the MLS2 sort? This is more plausible. Since the particles in the global population are distributed across many collectives, a causal process involving a number of collectives, such as selection, could conceivably affect the value of Cov (w_i, z_i), unlike that of Cov_k (w, z). How should such by-products be conceptualized? The most natural way is to do a contextual analysis in which particle fitness is the response variable, and collective fitness$_2$ is one of the independent variables:

$$w = \beta_1 z + \beta_2 Y + e$$

The idea here is to see if a particle's fitness is affected by the Y-value of the collective to which it belongs. If so, and if the particle character z correlates with Y, then a non-zero value of Cov (w_i, z_i) will be a side effect of direct selection on Y, at least in part. Since Y is collective fitness$_2$, this counts as a cross-level by-product of the MLS2 variety. The causal graph corresponding to this possibility is shown in Figure 3.9.

w (particle fitness)

Y
(collective fitness$_2$)

z
(particle character)

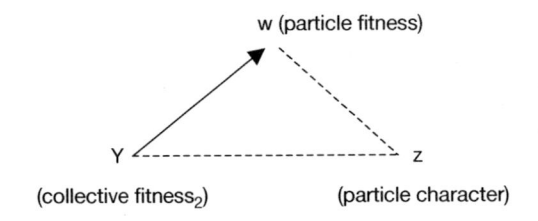

Figure 3.9. Collective→particle by-products in MLS2

This treatment of collective→particle by-products in MLS2 is analogous to their treatment in MLS1, discussed in Section 3.3.1 and illustrated in Figure 3.4; the only difference is that the contextual character is now collective fitness$_2$. As before, it is then straightforward to partition Cov (w_i, z_i) into two components, one of which reflects direct selection on z, the other a by-product of selection at the collective level.

How useful is this understanding of collective→particle by-products in MLS2? It might be thought unlikely that a particle's fitness will *ever* be causally affected by the fitness$_2$ of its collective, since these two fitnesses will often be measured over different timeframes. However, as noted in Chapter 2, we can always use a 'long-run' definition of particle fitness, that is, judge a particle's fitness by how many offspring it leaves after many generations, rather than one. So even if the collectives have a much slower turnover rate than the particles, a particle's long-run fitness may be affected by the fitness$_2$ of its host collective.

Eldredge and Gould (1972) discuss examples of by-products running from the species to the organism level which have essentially the above structure. In numerous mammalian clades, for example, horses, there is a trend towards increased body size over geological time, documented in the fossil record. Neo-Darwinists have traditionally attributed such trends to directional selection acting within species. But Eldredge and Gould argue that since most species are in 'stasis', the trend must result from selection *between* species. If larger horse species are fitter than smaller ones, that is, are more likely to survive and speciate, this could explain the trend in the absence of any within-species selection for large size. If this explanation is correct, then the fitness of any individual horse, judged by the number of descendents it leaves in the long run, is affected by the fitness$_2$ of its species, but is not affected by its own body size. Therefore, the global covariance between body size and organismic fitness in the clade as a whole, evidenced by the fossil record, is a by-product of selection at the species level—it disappears when we control for species fitness.

This suggests that our construal of collective→particle by-products in MLS2 is accurate. Of course, our construal says nothing about how common such by-products are likely to be. In a recent discussion, Gould (2002) appears to claim that such by-products will *always* exist. He writes: 'selection on higher-level individuals always sorts the lower-level individuals included within' (p. 28). Similarly, Vrba (1989) says that 'any selection process among individuals at a higher level will inevitably also dictate a component of sorting among included lower entities' (p. 139). But care is needed here. In a trivial sense the Gould/Vrba claim is true, for when a collective dies the particles within it must die too.

But it does not follow that the frequency of different particle *types* will change. As noted in Chapter 2, collective-level selection of the MLS2 sort permits no inference about the frequency of different particle types in the global population of particles, over the short or the long run (cf. Arnold and Fristrup 1982). So if the Gould/Vrba claim is interpreted as a claim about types, or about the ubiquity of collective→particle by-products as we have defined them, then it is untrue.

4

Philosophical Issues in the Levels-of-Selection Debate

INTRODUCTION

This chapter examines the philosophical literature on the levels of selection. Much philosophical discussion has focused on criteria for identifying the 'real' level(s) of selection in cases where there is ambiguity. Section 4.1 looks at two concepts, emergence and additivity, that have been regarded as important in this regard. Section 4.2 asks whether the concept of 'screening off', widely used in philosophical analyses of causality, provides any guidance. Section 4.3 looks at the doctrine known as 'pluralism', which says that in certain cases there *is* no objective fact about the level(s) at which selection is acting. Three possible arguments for pluralism, and their interrelations, are examined. Section 4.4 discusses reductionism in relation to the levels of selection.

4.1 EMERGENCE AND ADDITIVITY

4.1.1 The Emergent Character Requirement

Recall that in a multi-level scenario, where particles are nested in collectives, we can distinguish between emergent and aggregate characters of the collectives (cf. Chapter 2, Section 2.2.1). Many theorists have suggested that the aggregate/emergent distinction is somehow related to the levels-of-selection problem (Vrba 1989; Nunney 1993; Maynard Smith 1983). One version of this idea says that genuine collective-level selection, that is not reducible to selection at the particle level, can only operate on collective characters that are emergent. The intuitive force of the emergent character requirement is clear. Emergent characters are often complex, adaptive features of collectives, which it is hard to imagine evolving *except* by selection at the collective level. By contrast, aggregate characters are mere 'sums of the parts', so are natural candidates for arising from lower-level selection; certainly, they do not make convincing examples of collective adaptations.

One objection to this idea was noted in Chapter 2: aggregate characters often play a role in formal models of multi-level selection. This objection is not decisive, since formal models may not accurately reflect what happens in nature. But a deeper objection is that the emergent character requirement conflates product with process (cf. Williams 1992; Sober and Wilson 1998). The causal process of natural selection, at whatever level, must be distinguished from the products it can generate, such as complex adaptations. Emergent collective characters may be *evidence* of selection at the collective level, but should not be made preconditions of it; as Williams says, this is to put the cart before the horse (1992 p. 26).

This conclusion is bolstered by the analysis of particle→collective by-products in Chapter 3. That analysis aimed to make explicit the idea that a character-fitness covariance at the collective level might be a by-product of particle-level selection. The emergent character requirement can be thought of as a rival way of identifying particle→collective by-products; it says that the covariance at the collective level can only *not* be a by-product if the character in question is emergent. But our analysis implies that this is wrong. In both MLS1 and MLS2, whether a given character-fitness covariance is a cross-level by-product has nothing to do with whether the character is aggregate or emergent; these questions are independent in both directions. Knowing whether Z is aggregate or emergent tells us nothing about whether non-zero values of Cov (W, Z) or Cov (Y, Z) are by-products of lower-level selection, nor vice versa.

However, this does not close the case on emergent characters altogether. For as is clear from the causal graphs in Chapter 3, our analysis helps itself to the notion of a 'causal connection' between a character and fitness; it *uses* this notion to explain how cross-level by-products can arise in multi-level settings. It could be argued that where aggregate characters are concerned, it is impossible for there to exist a causal connection between the character and fitness, of either particles or collectives.[1] So the fact that the previous chapter's analysis deems the aggregate/emergent distinction irrelevant is beside the point; for it

---

[1] Lewontin (2000) makes a remark to this effect in the course of arguing that the average value of a trait in a group is not a real property of that group. He writes: 'averages are not inherited, they are not subject to natural selection; they are not physical causes of any events' (p. 33).

reaches that verdict by assuming that aggregate characters can be causally efficacious, which begs the question.

This is a coherent line of argument, but I do not think it is correct. The idea that a collective character has *got* to be emergent, if it is to causally influence the fitness of either the collective itself, or its constituent particles, does not seem plausible. If this were true, it would be a substantial metaphysical thesis in need of explanation. In general, the fitness of a biological unit can be affected by any of its characters, and by the environment. To single out some subset of these characters—the aggregate ones—and declare them incapable of causally affecting fitness is to claim a priori knowledge of what appears to be an empirical matter. Substantiating this claim would require a precise account of how to draw the aggregate/emergent distinction, and some indication of the source of the alleged difference in causal potential between the two sorts of character.

In Chapter 3, I suggested that the emergent character requirement stems from confusing two questions. The pertinent question is whether a given character-fitness covariance is a side effect of *selection* at a lower level, not whether it is a side effect of *some causal processes or other* at a lower level. If these two questions are not kept distinct, it becomes tempting to appeal to emergent characters to try to block the supervenience argument. Distinguishing the questions thus removes one possible motivation for thinking that genuine collective-level selection can only operate on emergent characters.

These considerations all suggest that the aggregate/emergent distinction is not of fundamental importance for the levels-of-selection question, a conclusion reached previously by Lloyd (1988), Grantham (1995), and Damuth and Heisler (1988). However, in one respect the emergent character requirement is on the right lines: it stems from a realization that a character-fitness covariance at a given level may not reflect direct selection at that level. This point is fundamentally correct, even though appealing to emergent characters is not the right way to accommodate it. Below I examine two further suggestions for how to identify the level(s) of selection that stem from the same realization; in a way, they constitute refinements of the emergent character requirement.

4.1.2 Additivity and the Wimsatt/Lloyd Approach

Wimsatt (1980) and Lloyd (1988) both argue that the concept of additivity holds the key to the levels-of-selection question. Additivity is another word for linearity. If two factors combine additively to produce a given effect, this means that the effect is a linear function of each factor's contribution. Less than perfect additivity means that the difference made by one factor depends on the other factor's contribution, so the factors interact. Non-additivity is one way that emergence might be made into a mathematically respectable notion.[2]

Additivity plays an important role in population genetics, where it is often important to know whether genes combine additively in the production of phenotypes, or whether they interact. Usually, geneticists talk not about additivity per se but additivity of *variance*.[3] Perfectly additive variance means that differences between entities, with respect to a given variable, are fully explained by differences in the independent contributions of some factors. So for example, if genotype fitness is the variable and the factors are two alleles A and B, the variance in genotype fitness is perfectly additive if and only if $(w_{AA} - w_{AB}) = (w_{AB} - w_{BB})$, that is, genotype fitness is a linear function of allelic 'dosage'. This means that fitness differences between genotypes are fully explained by differences in the number of A alleles they contain. The concept of additive variance is easily generalized beyond population-genetic contexts.

The idea that additivity might be relevant to the levels of selection is prima facie quite plausible. In a multi-level scenario, if all the variance in collective fitness is additive, then fitness differences between collectives are fully explained by differences in their particle composition; from which one might infer that particle-level selection is doing all the causal work. (This seems to be the intuition driving Wimsatt and Lloyd.) Interestingly, this suggestion ties in with Sewall Wright's (1980) approach to the levels question. For Wright, the distinction between 'genic' and 'organismic' selection, in the context of his 'shifting balance' theory of evolution, hinged precisely on additivity. Dominance and epistasis, which generate non-additive variance in genotype fitness, shift selection from the genic to the organismic level, according to Wright.[4]

---

[2] This suggestion has surfaced in the literature on occasion. Thus for example, Vrba (1984) defines an emergent character as one which is 'related by a nonadditive composition function to characters at lower levels' (p. 19).

[3] Strictly speaking, 'additivity' refers to the pattern of functional dependence itself, 'additive variance' to the pattern of statistical variation that the dependence gives rise to, so the concepts should not be equated. An effect could depend additively on two factors and yet show no additive variance, if the factors happen not to vary in the population. This is the basis of the distinction between 'statistical' and 'physiological' epistasis drawn by Wolf, Bradie, and Wade (eds.) (2000). Epistasis means non-additivity, in effect.

[4] However, Wright is using both 'genic selection' and 'organismic selection' in a non-standard sense. By the former, he means directional selection within a large panmictic

116   *Evolution and the Levels of Selection*

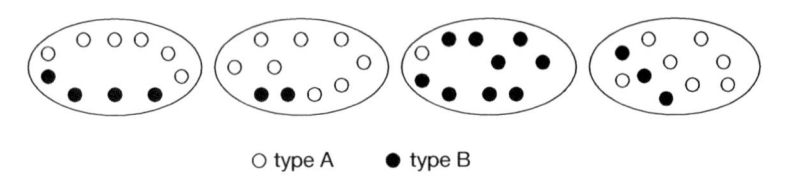

○ type A   ● type B

Figure 4.1. Particles of two types nested within collectives

More recently, Michod (1999) has argued that non-linear interactions play a key role in transitions between units of selection.

Despite its prima facie plausibility, the Wimsatt/Lloyd idea has been heavily criticized in the literature (Lewontin 1991; Godfrey-Smith 1992; Sarkar 1994; Sober and Wilson 1994). My strategy will be to use the foregoing analysis of cross-level by-products to try to adjudicate the debate.

Consider a multi-level scenario with strict nesting and a fixed number of particles per collective, as before. Particles are of two types, A and B, found in differing proportions in different collectives (Figure 4.1). Consider a collective character Z which varies in the population, that is, Var $(Z) \neq 0$. It is possible that a collective's Z-value depends on the proportion of A-particles that it contains. If the dependence is perfectly linear, as in Figure 4.2A, then all the variance in Z is additive. This means that differences in collectives' Z-values are fully explained by the differing proportions of A-types that they contain. A non-linear pattern of dependence is shown in Figure 4.2B. This means that at least some of the variance in Z is non-additive. Note that we can take Z to be fitness itself—either fitness$_1$ or fitness$_2$—permitting us to ask whether all the variance in collective fitness is additive.

The simplest version of the Wimsatt/Lloyd proposal is that if all the variance in fitness at a level is additive, then selection is not acting at that level—all the action is at a lower level.[5] More precisely, I will take the additivity proposal to be the conjunction of two claims:

(I) if there *is* collective-level selection, there must be non-additive variance in collective fitness, and

---

population of the sort associated with Fisher (1930). By the latter, he means diffusion of co-adapted genotypes between partially isolated demes of a species. Note also that Wright's original presentation of the shifting-balance theory does not use the terminology of genic and organismic selection; see for example Wright (1931).

[5] This formulation is closer to Wimsatt (1980) than to Lloyd (1988).

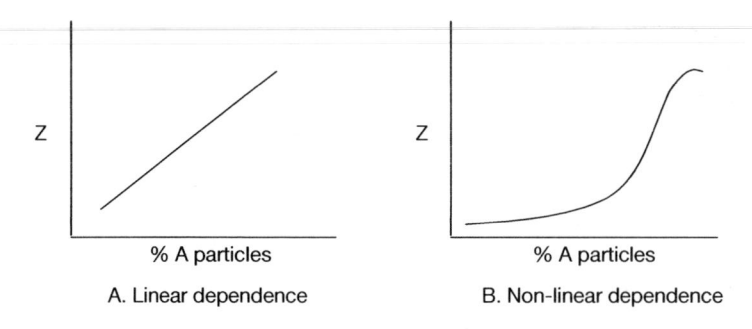

Figure 4.2. Dependence of collective character Z on particle composition

(II) if there is *no* collective-level selection, then any variance in collective fitness must be additive.

In effect, propositions (I) and (II) constitute a rival to the account of cross-level by-products in Chapter 3. Note that the status of (I) and (II) may differ depending on whether 'collective fitness' is defined as fitness$_1$ or fitness$_2$, that is, on whether we are dealing with MLS1 or MLS2.

Sober and Wilson (1994) argue that additivity is irrelevant to the levels question, on the basis of a simple multi-level model for the evolution of altruism of the MLS1 sort. Organisms of two types, altruists (A) and selfish (B), are distributed into groups in varying proportions. Within each group, individual selection favours the selfish types; but group-level selection favours groups with a high proportion of altruists. Sober and Wilson note that in this model, it is quite possible for group fitness to be a linear function of proportion of altruists.[6] If so, then all the variance in group fitness is additive—fitness differences between groups are fully explained by their differing proportions of A-types. Group selection is thus compatible with perfectly additive variance in group fitness; so proposition (I) is false.

What about proposition (II)? Sober and Wilson's example seems to show that it is false too. In their model, there are only two ways in which there can be no group selection—all the groups must have equal fitness, or group fitness and group character must be uncorrelated. If the former, then there *is* no variance in group fitness, additive or otherwise. If the latter, it does not follow that all the variance in group fitness is additive. On the contrary, if group fitness and group character (proportion of

---

[6] Linear fitness functions are commonly used in models of this sort; see for example Hamilton (1975), Wilson (1990), or Kerr and Godfrey-Smith (2002).

A-types) are uncorrelated, then group fitness *cannot* depend linearly on proportion of A-types, so all the variance in group fitness *cannot* be additive. Hence (II) is false too.

This *appears* to show that in MLS1, additivity is wholly irrelevant to determining the levels of selection. Group-level selection does not require non-additive variance in group fitness, *pace* (I); and the absence of group-level selection does not require perfectly additive variance, *pace* (II). However, there is a complication. For Sober and Wilson's argument implicitly relies on the Price approach: group selection is taken to mean a covariance between group fitness and group character. But as we saw in Chapter 3, this covariance may be a by-product of lower-level selection. This possibility led us to explore contextual analysis, which constitutes a rival to the Price approach. What happens when we view the additivity proposal through the lens of contextual analysis?

It is easy to see that proposition (I) comes out false, on the contextual approach. For recall that under pure 'soft selection', where all the collectives have identical fitness, contextual analysis says that there is nonetheless a component of selection at the collective level. So on the contextual approach, collective-level selection does not require *any* variance in collective fitness, let alone non-additive variance.

Matters are different with proposition (II). On the contextual approach, the absence of collective-level selection means the absence of 'collective effects' on particle fitness—a particle's fitness depends only on its own character. This means that any two A-particles have identical fitness whichever collective they live in, and similarly for B-particles; we can denote these by w_A and w_B. It immediately follows that any variance in collective fitness *is* additive, for collective fitness must be a linear function of proportion of A-types. That is, if $W(x)$ is the fitness of a collective containing x A-types and (n-x) B-types, where n is collective size, then $W(x)$ is linear in x—for $(W(x+1) - W(x)) = (w_A - w_B)$, which is constant by hypothesis. So on the contextual approach, the absence of collective-level selection implies that all the variance in collective fitness$_1$ is additive—just as proposition (II) asserts.

This means that the relevance of additivity depends on whether we favour the Price or the contextual approach to MLS1. On the Price approach, additivity is wholly irrelevant to determining the levels of selection, as Sober and Wilson's example shows. On the contextual approach, additivity is partially relevant—one of the two propositions that we have identified with the additivity proposal comes out true. The previous chapter argued that the contextual approach is theoretically preferable to

Philosophical Issues in the Levels-of-Selection Debate 119

the Price approach, on balance. I suggest that this constitutes a partial vindication of Wimsatt's original intuition.

What about MLS2? Recall that in MLS2, it proved difficult to identify a sure-fire criterion for genuine collective-level selection, that is not the result of a cross-level by-product; we succeeded only in identifying a *sufficient* condition for all the character-fitness covariance at the collective level to be a by-product of particle-level selection. Therefore, the status of propositions (I) and (II) cannot be conclusively decided when 'collective fitness' is interpreted as fitness$_2$; this represents an unsolved problem.

4.1.3 Emergent Relations and the Damuth–Heisler Approach

In two papers on multi-level selection, Heisler and Damuth (1987) and Damuth and Heisler (1988) provide an insightful discussion of emergence, and link it to the contextual approach to MLS1. They argue that emergence *is* relevant to the levels of selection, but not in the way that advocates of the emergent character requirement have thought. The crucial question is not whether a given character is emergent rather than aggregate, but whether the *relation* between the character and fitness is emergent, Damuth and Heisler argue.

The idea of a relation being emergent may sound unusual, but there is no reason why it should not make sense, at least to the extent that emergent characters make sense. For often in metaphysics, a distinction between types of property will be matched by an analogous distinction between types of relation. Damuth and Heisler say that a character-fitness relation is emergent if it 'cannot be accounted for' by a character-fitness relation at a lower hierarchical level. They argue that contextual analysis provides a way of identifying such emergent relations, in the MLS1 case. Since our treatment of cross-level by-products in MLS1 was also based on contextual analysis, it ties in with Damuth and Heisler's notion of an emergent relation.

To see this, recall that contextual analysis asks whether a particle's fitness depends only on its own character, or is also affected by a collective character (aggregate or emergent). If there is a 'collective effect' on particle fitness, then the statistical association between collective character and fitness remains after controlling for particle character. In Damuth and Heisler's terms, this means that the relation between collective character and fitness is emergent, because it cannot be fully explained by the relation between particle character and fitness. By contrast, where a particle's fitness depends only on its own character, the relation between collective character and fitness (if any) is not emergent, for it is a side effect of lower-level selection; in our terms, there is a particle→collective by-product. Understood this way, the relevance of emergent relations to the levels-of-selection question is immediate.

One consequence of Damuth and Heisler's definition is that 'emergent relation' actually becomes a *better* defined notion than emergent character. For recall that contextual analysis treats all collective characters equally—they are all just possible sources of causal influence on particle fitness. So on the contextual approach, the distinction between aggregate and emergent characters is of no particular significance. A theorist who held that this distinction is not a principled one, or cannot be satisfactorily explicated, could still accept Damuth and Heisler's account of when a given character-fitness relation is emergent. For their account does not presume that the notion of emergent character is antecedently understood.

Interestingly, Damuth and Heisler restrict their discussion of emergent relations to multi-level selection of the MLS1 variety.[7] But there seems no reason why it cannot be extended to MLS2. For cross-level by-products can also arise in MLS2; putative examples from the literature were discussed in Chapter 3. Since Damuth and Heisler's concept of an emergent relation is intimately linked to the concept of a cross-level by-product, this suggests that emergent character-fitness relations can also occur in MLS2. The only way to deny this would be to argue that cross-level by-products cannot occur in an MLS2 framework, which is implausible. For this would be to say that, in principle, natural selection at the particle level *cannot* be the explanation for why collective character and fitness$_2$ covary; but this is surely an empirical issue. Admittedly, cross-level by-products in MLS2 are difficult to analyse, for the reasons explained previously, but we should not deny their existence.

In a recent discussion, Gould (2002) describes what he calls an 'emergent fitness' approach to multi-level selection; he contrasts this with the 'emergent character' approach. The same contrast is drawn by Lloyd and Gould (1993), and Gould and Lloyd (1999). The inspiration for the 'emergent fitness' idea comes from Damuth and Heisler, though Gould and Lloyd's concern is with multi-level selection of the MLS2 sort. This is not in itself a problem, if I am right that Damuth and Heisler's notion

---

[7] Damuth (personal communication) argues that the concept of an emergent relation is not well-defined in MLS2, and would not be of importance even if it were well-defined.

Philosophical Issues in the Levels-of-Selection Debate 121

of an emergent character-fitness relation can be extended from MLS1 to MLS2. However, in one place Gould states that his emergent fitness criterion, as applied to species selection, implies that genuine species selection occurs wherever there is a species character such that 'the fitness of the species covaries with the character' (2002 p. 661). Lloyd and Gould (1993) make a similar remark (p. 598). But this is clearly wrong. The whole point of the notion of an emergent relation, for Damuth and Heisler, is to take account of the possibility that character-fitness covariance at a level might *not* be due to direct selection at that level. If Gould's description of what his 'emergent fitness' criterion amounts to is taken at face value, then the criterion is incorrect.

4.2 SCREENING OFF AND THE LEVELS OF SELECTION

In a series of publications, R. Brandon has argued that the concept of *screening off* holds the key to the levels-of-selection question (Brandon 1982, 1988, 1990; Brandon et al. 1994). Screening off was introduced into philosophy by Reichenbach (1956) as a way of trying to analyse causality in terms of probability; Salmon (1971) developed this idea further. Given three events A, B, and C, A is said to screen off B from C if conditionalizing on A renders B and C probabilistically independent, that is, if $P(C/A$ and $B) = P(C/A)$. This means that if you already know whether event A occurs, information about whether B occurs does not affect the probability that C occurs.

Brandon uses screening off in the service of two projects. The first is to help explicate the claim, made by Mayr (1963) and Gould (1984), that organisms' phenotypes but not their genotypes are directly 'visible' to natural selection.[8] The second is to provide a general account of the conditions required for selection to act at a given level. Though Brandon apparently sees a link between these two projects, I follow Walton (1991) and Sober (1992) in regarding them as unrelated. Here I focus on Brandon's second project, leaving aside his use of screening off to unpack the Mayr/Gould idea.

The idea that screening off might help identify the levels of selection is quite plausible. Indeed, our analysis of cross-level by-products in

---

[8] This 'visibility' argument was an early objection to the 'genic selectionism' of Williams (1966) and Dawkins (1976).

Chapter 3 in effect used screening off—for the concept of screening off is closely related to the statistical concept of controlling for a variable, that is, of partial correlation/regression, which we used to explain how a character-fitness covariance at one level could be caused by selection at another level.[9] Therefore, it should be easy to find a logical link between Brandon's analysis and our own.

Brandon's starting point is that for selection to act at a given level, it is not sufficient that entities at that level should vary in fitness. What is needed, in addition, is that the fitness differences have the right sort of explanation. The additional requirement is that 'the "phenotypes" of entities at that level [must] screen off properties of entities at every other level from reproductive values at the given level' (1990 p. 88). So in a two-level scenario, Brandon's proposal says that for selection to act at the level of the collectives, collective character must screen off all particle characters from collective fitness. This is meant to capture the idea that a collective's fitness depends on its *own* character, rather than on the characters of its constituent particles.

Note that this proposal could be applied to either MLS1 or MLS2, depending on whether 'collective fitness' is interpreted as fitness$_1$ or fitness$_2$. However, Brandon confines his attention to MLS2, so for the moment I do likewise.

Brandon's proposal can be expressed more precisely, as follows.[10] As before, let Z and Y denote collective character and fitness$_2$ respectively; let $z_1 \ldots z_r$ be an exhaustive list of particle characters. For selection to act at the collective level, the following must be true, according to Brandon:

$$E(Y/Z \text{ and } z_1 \ldots z_r) = E(Y/Z)$$
$$E(Y/Z \text{ and } z_1 \ldots z_r) \neq E(Y/z_1 \ldots z_r)$$

where E(A/B) denotes the conditional expected value of A on B. The first statement says that the expected value of Y, conditional on Z, is the same as its expected value conditional on both Z and $z_1 \ldots z_r$. This captures the idea that once collective character is taken into

---

[9] See Irzik and Meyer (1987) for a good discussion of the relation between screening off and partial correlation.

[10] This formalization of Brandon's proposal is based on one given by Sober and Wilson (1994). It differs slightly from Brandon's own wording, in that it uses conditional expected values rather than conditional probabilities to express screening off. This is the obvious way to generalize the concept of screening off to non-dichotomous random variables.

account, particle characters convey no additional information about collective fitness. The second statement says that the expected value of Y, conditional on both Z and $z_1 \ldots z_r$, differs from its expected value conditional on $z_1 \ldots z_r$ alone. This captures the idea that collective character does convey information about collective fitness, even when particle characters are taken into account. So taken together, the two statements say that there is an asymmetry between Z and $z_1 \ldots z_r$ with respect to their relevance to Y.

Sober and Wilson (1994) argue that Brandon's criterion for collective-level selection can *never* be satisfied, because any collective character must be determined by the characters (intrinsic and relational) of its constituent particles.[11] It cannot be the case that once all particle characters are taken into account, additional information about collective character alters the collective's expected fitness; for the particle characters determine the collective character. Sober and Wilson write: 'we are puzzled as to why the inequality demanded by [Brandon's] criterion should ever be true, since the unary and relational properties of individuals evidently determine the properties of the group' (1994 p. 547). (Read 'particle' and 'collective' for 'individual' and 'group' in this quotation.) So on plausible metaphysical assumptions, Brandon's criterion for higher-level selection is unsatisfiable.

It is useful to compare Brandon's analysis with our own. Recall that in Chapter 3, we identified a *sufficient* condition for the existence of a cross-level by-product in MLS2, namely if differences in collective fitness$_2$ are fully explained by differences in average particle fitness, rather than by differences in collective character Z. Expressed in terms of screening off, this condition says that for selection at the collective level, Z must screen off *average particle fitness* from collective fitness$_2$; otherwise, all the selection is at the particle level. By contrast, Brandon requires that Z must screen off *all particle characters* from collective fitness$_2$. Both conditions are meant to capture the idea that fitness differences between collectives must result from differences in Z, rather than being 'caused from below'. But the conditions differ with respect to *which* lower-level characters must be screened off by Z, for selection at the collective level.

Our condition is preferable to Brandon's for two reasons. First, it is immune from the Sober and Wilson objection. For average particle fitness clearly does *not* determine every collective character; so it is quite

---

[11] Though see Chapter 2 footnote 12 for qualifications to this claim of determination.

possible for Z to be positively relevant to collective fitness even after conditioning on average particle fitness. Secondly, we gave an argument for *why* average particle fitness is the property that must be screened off by Z, for genuine collective-level selection. The argument was that if average particle fitness is *not* screened off by Z, then fitness differences among the particles are ultimately responsible for the character-fitness covariance at the collective level, so the latter is spurious. Thus our condition, unlike Brandon's, is both logically satisfiable and supported by a theoretical argument.

At root, Brandon goes wrong because he runs together the question whether *selection* at the lower level is responsible for the fitness variation at the higher level, with the question whether *some lower-level causal process or other* is responsible. Previously we argued that the former, not the latter, is the relevant question in the context of the levels-of-selection debate. If the two questions are not kept separate then the supervenience argument beckons—and threatens to make genuine higher-level selection impossible. Sober and Wilson's observation that Brandon's screening off criterion cannot be satisfied, given the assumption of part–whole supervenience, neatly illustrates this point.

Finally, although their criticism of Brandon is correct, Sober and Wilson (1994) imply that screening off is of no use at all for identifying the levels of selection. But this is not so, if the analysis of the previous chapter is right. The possibility that a character-fitness covariance at one level is a by-product of selection at another level lies at the heart of the levels-of-selection problem; and the natural way of explaining what this possibility amounts to makes use of screening off, or closely related concepts. (More precisely, cross-level by-products were explained in terms of patterns of causal connection between variables that will induce the screening off relation.) Brandon is not wrong to think that screening off is relevant; his mistake lies only in his account of *what* must be screened off, for genuine collective-level selection to occur.

It is not difficult to see why Sober and Wilson regard screening off as irrelevant. Their interest lies primarily in multi-level selection of the MLS1 sort, and they conceptualize MLS1 using the Price approach— particle-level selection operates on fitness differences *between* particles *within* collectives, and collective-level selection operates on fitness differences between the collectives themselves.[12] The Price approach to MLS1

---

[12] Sober and Wilson (1994) do not directly discuss Price, but Sober and Wilson (1998) do; the latter work makes clear that their conceptualization of multi-level selection

is heuristically valuable, as we have seen, but arguably it is theoretically flawed, for it fails to take account of cross-level by-products. Since screening off enters the picture precisely when we try to understand such by-products, this explains why Sober and Wilson do not appreciate its relevance.

4.3 REALISM VERSUS PLURALISM ABOUT THE LEVELS OF SELECTION

The issue of realism versus pluralism is a recurring theme in philosophical discussions of the levels of selection. Realists hold that the question 'at what level or levels is selection acting?' always has a uniquely correct answer; pluralists deny this. Typically, pluralists argue that certain selection processes can legitimately be described in more than one way; the choice between the competing descriptions is conventional, not factual, they claim.

The realism/pluralism issue has arisen in relation to many different areas of the levels discussion. Examples include the debate between 'genic' and 'organismic' selectionists prompted by Dawkins' work; the debate over how 'inclusive fitness' models should be conceptualized; and the debate over 'trait-group' selection. These debates are explored in subsequent chapters. But it would be useful if something general could be said about realism and pluralism, to provide a touchstone for assessing specific cases. First, some conceptual clarifications are necessary.

One very weak sense of pluralism is just the idea that selection *can* occur at different levels, so the actual level(s) of selection must be assessed on a case-by-case basis. Sober and Wilson (2002a) describe this as 'pluralism concerning what happens in nature' (p. 529). This type of pluralism is uncontroversial, and does not conflict with realism. Acknowledging that selection *can* occur at different levels is compatible with saying that in any given case, it is an objective fact which level or levels it *is* occurring at.

---

is faithful to the Price approach. However, there is a slight exegetical complication. For in a footnote, Sober and Wilson (1998) say that contextual analysis is actually superior, but that the difference between the Price and contextual approaches is unimportant for their purposes (p. 343n). However, Sober and Wilson (personal communication) both insist that in the case of pure soft selection discussed in Chapter 3, which is the critical test between the Price and contextual approaches to MLS1, the former gets the answer right; furthermore, Sober and Wilson (2002c) explicitly endorse the Price approach.

Another weak sense of pluralism says that it is a mistake to enquire about *the* level of selection, because selection can occur simultaneously at multiple levels. Pluralism of this sort is also uncontroversial. However, in early philosophical discussions there was a tendency to try to classify problem cases as *either* 'group selection' *or* 'individual selection' *or* 'genic selection' and so on—as if the operation of one of these modes of selection excludes the others. 'Monism' is an appropriate label for the idea.[13] Monism must obviously be rejected, for it is tantamount to denying the very possibility of multi-level selection. But rejecting monism does not mean rejecting realism, so is philosophically innocuous.

A more interesting form of pluralism says there is 'no fact of the matter' about the true level(s) of selection, so the choice between the competing alternatives is conventional. For example, a pluralist might say that attributing a trait's evolution to individual selection and attributing it to a combination of individual and group selection are both legitimate; there is no objective fact about which attribution is correct. This thesis *is* philosophically interesting, and does conflict with realism; it is how pluralism will be understood here.[14]

Pluralist theses can be local or global in scope. A local version would say that in certain specific cases, there is no objective fact about the level(s) of selection; a global version would say that there is never such a fact. Both versions are found in the literature. Sterelny and Kitcher (1988) defend a global pluralism, as does Kitcher (2004); on his view, it involves a kind of metaphysical mistake to think in terms of the 'true' level(s) of selection.[15] By contrast, Dugatkin and Reeve (1994) defend a local pluralism. They hold that pluralism is true of a certain class of evolutionary processes, for reasons specific to that class. Clearly, arguments for local and global pluralism must be different in kind.

Pre-theoretically, realism seems more plausible than pluralism. Biologists who have disagreed over the levels of selection often write as if their disagreement were straightforwardly factual. For example, when Wynne-Edwards (1962) and Lack (1966) disagreed over whether population self-regulation in birds had evolved by group selection, they

[13] Though note that Kitcher, Sterelny, and Waters (1990) and Barrett and Godfrey-Smith (2002) both use 'monism' in a somewhat different sense.

[14] Useful discussions of the concept of pluralism in relation to the levels of selection include Barrett and Godfrey-Smith (2002), Sober and Wilson (2002a, c), and Godfrey-Smith and Kerr (2002).

[15] Kitcher traces this mistake to the 'realist sympathies' of certain philosophers, which have led them to take the 'metaphor' of natural selection too seriously (2004 p. 90).

took themselves to be debating a factual question about the course of evolutionary history. To suggest that there is no objective fact about who was right, or that their competing claims were 'equally valid perspectives on the facts', seems simply wrong.

Previously we distinguished the qualitative question 'at what levels is selection occurring?' from the quantitative question 'how *much* of the total evolutionary change is due to selection at each level?' In the recent literature on multi-level selection, realist answers to both questions have been presumed. The idea that the total evolutionary change can be partitioned into components corresponding to different levels of selection lies at the heart of the Price approach to multi-level selection, and the modifications of that approach such as contextual analysis. Such a causal decomposition is only possible if realism is correct.

The fact that realism is implicit in most biological work is an important consideration, but is not decisive. For scientific theorizing does sometimes rest on questionable metaphysical assumptions. And it would be wrong to imply that pluralism is 'just for philosophers'. Biologists with pluralist sympathies include Buss (1987), whose ideas have inspired much of the recent work on evolutionary transitions. Buss himself favours a multi-level approach to selection, but argues that the empirical facts can also be captured by a single-level approach. He writes: 'the choice between... a single unit of selection [and]... multiple units of selection, is a choice of *language*' (p. 177, my emphasis). Other biologists with pluralist sympathies include Dugatkin and Reeve (1994).

The abstract treatment of multi-level selection developed earlier embodies a realist outlook. Selection occurs at a given level, we argued, if entities at that level vary with respect to a character which causally influences fitness. Whether such a causal relation obtains in any case was assumed to be an objective matter. Our account of cross-level by-products retained this realist bias. That account tried to identify the conditions under which a character-fitness covariance at one level might arise from direct selection at another; the obtaining of those conditions was taken to be an objective matter. So if the framework developed so far is correct, it is hard to see how there can be *any* room for pluralism, local or global. So where do pluralistic ideas come from, and why does our framework apparently exclude them?

I think there are three sources, each of which involves rejecting an implicit assumption of our framework. The first has to do with causation, the second with the reality of the biological hierarchy, the third with mathematically interchangeable descriptions. I explore them in turn.

4.3.1 Pluralism and Causality

The simplest way of compromising the realism implicit in our framework is to adopt a non-realist view of causation. On such a view, statements of the form 'X causally affects Y' do not report objective facts about the world at all. Anyone holding such a view would deny that it is an objective matter whether a given character causally affects fitness. Similarly, if one held that causation is nothing but correlation, one would deny the reality of the distinction between character-fitness covariance that does and does not reflect direct selection. Given such metaphysical views, the levels-of-selection question can at most be a question about the heuristic utility of one description over another.

Non-realist views of causation, and a refusal to accept the causation/correlation distinction, have a long precedent among empiricist philosophers and scientists, but they enjoy little popularity today.[16] Few if any writers on the levels of selection explicitly endorse non-realism about causation. Nonetheless, it is sometimes suggested that to talk about the 'true' levels of selection involves a mistaken reification of causal relations, or ignores the fact that causal chains can be chopped up in multiple ways. For example, Kitcher (2004) argues that it makes no sense to ask about 'the *real* locus of causation in selection processes' (p. 89, emphasis in original). He writes: 'one can tell all the facts about how genotype and phenotype frequencies change across the generations—including the causal explanations of the changes—without any commitment to a definite level at which selection acts' (ibid. p. 89). Far from being a natural default position, realism about the levels of selection is the result of inflationary metaphysics, according to Kitcher.

This argument seems to me unconvincing. To ask about the 'real locus of causation' is to ask about the hierarchical level or levels at which character differences cause fitness differences. So presuming we are realists about questions of the form 'does X causally affect Y?', Kitcher's assertion seems wrong. If the explanation for why a given particle character has changed in frequency is that it causally affects the fitness of particles, for example, then selection has acted at the particle level; if not, then not. Given this conception of the levels-of-selection question, it is clearly *not* possible to give causal explanations of evolutionary changes 'without any commitment to a definite level at which selection acts'.

---

[16] For example, Hume, Mach, Pearson, and Russell all defended such views. See Price (2006) for a recent, sophisticated version of non-realism about causation.

For different causal explanations embody different such commitments, and thus conflict.

However, this may be slightly too quick. From the context of the above quotation, Kitcher is alluding to the version of pluralism associated with 'genic selectionism'. As usually formulated, this view does not say that the level of selection is a wholly indeterminate matter, but rather than any selection process, at whatever level, can *also* be described as a case of genic selection (Dawkins 1982; Buss 1987; Sterelny and Kitcher 1988). So some levels-of-selection disputes do have objectively correct answers; it is only those disputes in which one of competing levels is the genic level that receive a pluralistic resolution. This view is examined in Chapter 5; properly understood, it turns out to be *compatible* with the realism implicit in our framework.

One might think that a causality-based argument for pluralism can be extracted from the concept of a cross-level by-product. Since selection at one level can have effects that filter up or down the hierarchy, perhaps the notion of the real level(s) of selection, or the real 'locus of causation' loses its grip? But this argument is incorrect, for it conflates direct and indirect selection. The filtering of effects from one level to another does not imply that there is no fact about whether any given character-fitness covariance is the result of a direct causal link. Cross-level by-products make it hard to discover the causal basis of evolutionary change, but do not undermine the idea that there is one true causal story to be told.

Finally, another possible causality-based route to pluralism must be eliminated. Recall the supervenience argument, which says that *any* higher-level character-fitness covariance must be spurious, because collective properties supervene on particle ones, so the causal arrow is always at the lower level. One could respond to this argument in a pluralistic way, that is, by saying that there is no objective fact about what level the causal arrow is at, or that causal arrows at both levels can peacefully coexist. In the philosophy of mind, these responses to the supervenience argument have sometimes been touted.[17] If they work, they presumably must apply in biology too. But the resulting pluralism would not be pluralism about the levels of *selection*. For as stressed previously, the lower-level causal story need not be a selective story. So even if the choice

---

[17] The idea that causal arrows at both levels can peacefully coexist is sometimes called 'causal overdetermination'. Loewer (2002) defends this idea as a way of responding to the supervenience argument developed by Kim (1998), in the context of the mind/body problem.

between lower- and higher-level causal stories is conventional, this does not make the level of *selection* a matter of convention.

4.3.2 Pluralism and Hierarchical Organization

A quite different motivation for pluralism stems from worries about the reality of the biological hierarchy itself. Our abstract treatment of multi-level selection assumes the existence of part–whole structure. But disagreements over such structure are possible. For example, some theorists regard social bacteria colonies as genuine wholes; others regard them as mere aggregates, lacking true individuality (cf. Szathmáry and Wolpert 2003; Shapiro and Dworkin 1997). Analogous disagreements surround the status of certain animal groups (cf. Maynard Smith 2002; Sterelny and Griffiths 1999). One could attempt a pluralist resolution of these disagreements, arguing that there is no objective fact about who is right. This would lead naturally to pluralism about the levels of selection. If it is indeterminate whether bacterial colonies are real entities, it may also be indeterminate whether colony-level selection occurs.

Obviously, this sort of pluralism will be local rather than global, for many instances of part–whole structure are unproblematic. Modern metazoans, for example, are clearly real entities, as are ant and termite colonies. So whether selection occurs at these levels will be an objective matter. Pluralism will only get a hold in the grey area between 'genuine' collectives and mere aggregates of lower-level particles. The existence of such grey areas is unsurprising, for most binary distinctions are vague at the edges. Such vagueness does not impugn the reality of a distinction, so long as there are clear-cut cases on either side.

In the present context something stronger can be said. Since the biological hierarchy is *itself* the product of evolution, the existence of grey areas is virtually guaranteed. For example, multicelled organisms evolved from single-celled ancestors, through a series of intermediate stages. Whatever we take to be the defining features of true multicellularity, it is practically certain that those features evolved gradually. So even if we knew all the intermediate stages, we could not identify a sharp cut-off point signalling the advent of the first multicelled creatures. Clearly, the same goes for entities at other hierarchical levels too. Note that this does not imply that currently existing problem cases, for example, bacterial colonies, represent transitional stages en route to 'true' wholes, though this is possible.

Philosophical Issues in the Levels-of-Selection Debate 131

How significant is pluralism stemming from this source? It may seem uninteresting—just the old philosophical problem of vagueness in biological guise. But some commentators see it as quite significant. For example, Sterelny and Griffiths (1999) describe a range of 'collective individuals' whose status as genuine entities is in doubt, including baboon troops, wolf packs, and beaver families (p. 166–77). The indeterminate status of these entities means there is no objective fact about whether selection acts on the collectives themselves or only on their constituent individuals, they argue; both are valid perspectives on the facts.

Interestingly, Sterelny and Griffiths link their pluralism about levels of selection with a non-realist thesis about the organism/environment boundary. Genuine collectives, such as ant colonies and multicelled organisms, are separated from their environment by a physical boundary in a way that 'collective individuals' are not, they claim. A wolf pack can be regarded as an entity in its own right, but can alternatively be regarded as part of the social environment of the individual wolves. So by drawing the organism/environment boundary at different points, we can switch between thinking of the particles and the collectives as the units of selection. In effect, this means that collective-level selection can be re-conceptualized as particle-level selection in a more complex social environment.

It might be objected that Sterelny and Griffiths's pluralism conflates process with product. Certainly, some collectives are more cohesive than others, but this is a *consequence* of collective-level selection, not a criterion for it. Cohesion and physical boundedness are adaptations of collectives that evolved gradually, so should not be made preconditions of genuine collective-level selection—this leads to what we called the 'chicken-and-egg problem' in Chapter 2. So Sterelny and Griffiths are wrong to base a taxonomy of selective processes on features that are themselves products of selection.

This criticism is partly right, but slightly unfair. For Sterelny and Griffiths's pluralism can be read as a way of *addressing* the chicken-and-egg problem, rather than as something that falls prey to it. Recall that in Chapter 2, we handled the chicken-and-egg problem by adopting an extremely liberal conception of part–whole structure, according to which any amount of interaction between particles is sufficient to bring a collective into existence. This avoids the chicken-and-egg problem, and renders the levels-of-selection question fully objective—collective-level selection either occurs or it does not. But this objectivity is an artefact of using the interactionist definition of collective to impose a sharp boundary where, arguably, none exists. Sterelny and Griffiths's suggestion that it is sometimes indeterminate whether real collectives exist, and thus whether there is selection at the collective level, avoids the imposition of a sharp boundary, and so represents an alternative to the interactionist conception.

Seen in this light, Sterelny and Griffiths's pluralism is a way of coming to grips with the fact that an unbroken lineage leads from 'cohesive' collectives such as multicelled organisms and ant colonies, to loose aggregates of interacting particles, to single particles with a free-living existence. The interactionist conception also comes to grips with this fact, in a way that preserves realism about the levels of selection, but at the price of a somewhat arbitrary stipulation about what constitutes a collective. Sterelny and Griffiths avoid such a stipulation, but at the price of sacrificing realism.

Is there any reason to prefer one approach over the other? One argument we used for the interactionist account is that it tallies well with a standard conception of multi-level selection, according to which higher levels of selection arise via fitness-affecting interactions between lower-level units. But the argument is not conclusive; opponents of this conception of multi-level selection would simply reject the interactionist account of part–whole structure too. Moreover, Sterelny and Griffiths can also accommodate the idea that interactions give rise to higher levels of selection; they just deny that we *have* to recognize two levels of selection as soon as there are fitness-affecting interactions among the particles.

If this is right, then Sterelny and Griffiths's pluralism constitutes a legitimate alternative to the interactionist conception of a collective. But it raises a number of questions. Can all types of higher-level selection be re-conceptualized as lower-level selection in a structured environment? Are there circumstances where this re-conceptualization is impossible? And how do we integrate 'pluralism' with a formal description of the evolutionary dynamics? These questions are addressed in the next section.

To summarize so far: Realism about the levels of selection is the natural default option, is implicit in most but not all biological work, and is embodied in the abstract account of multi-level selection developed previously. Two possible sources of pluralism are (i) non-realism about causality, and (ii) indeterminacy of hierarchical structure. The first of these is not especially plausible, and plays at most an indirect role in recent discussions. The second has some importance, particularly in the context of evolutionary transitions where borderline cases of part–whole structure are inevitable. It argues for a local rather than a global sort of pluralism.

4.3.3 Pluralism and Multiple Representations

The third source of pluralism arises from the fact that a single evolutionary process can often be modelled, or mathematically represented, in different ways. In itself, this is unsurprising. Models in science provide an idealized description of reality, focusing on some features at the expense of others. So constructing a model requires making a choice about which features to leave out; different choices will lead to alternative representations of the same event or process.

Importantly, the existence of multiple representations only leads to an interesting form of pluralism if the representations in question are prima facie incompatible. If not, then there is no clash with realism—both representations can be accepted as correct descriptions of the world. The equivalence of the Heisenberg and Schrödinger formulations of quantum mechanics, though surprising when first demonstrated in the 1930s, did not threaten realism, because the formulations, though different, were not prima facie incompatible.[18] Physicists responded to their proven equivalence not by arguing that there is no objective fact about which is correct, but simply by accepting both.

Multiple representations play a role in many levels-of-selection debates. A single evolutionary process can sometimes be modelled in two different ways, with seemingly incompatible implications about the level(s) of selection. For example, the first might treat fitness as a property of particles alone, while the second might attribute fitness to collectives, or to both particles and collectives. On the face of it these representations conflict, making competing claims about the level(s) at which natural selection is acting. If both representations can capture the evolutionary dynamics equally well, then a pluralist conclusion beckons. Perhaps there is no fact of the matter about which representation is correct, nor therefore about the 'real' level(s) of selection? Arguments of this sort are used by Maynard Smith (1987a), Dugatkin and Reeve (1994), Sterelny and Griffiths (1999), Kerr and Godfrey-Smith (2002), R. A. Wilson (2003), and others.

Pluralism is not the only possible response to the existence of multiple representations. An alternative is to argue that the two representations

---

[18] Though quantum mechanics does threaten realism for other reasons.

are not equally good, for example, because one is conceptually more basic, or more informative, or more faithful to the causal structure of the evolutionary process being modelled. This last criterion—fidelity to causal structure—is often invoked by opponents of pluralism. Typically, they argue that a representation may correctly predict the evolutionary dynamics without capturing the underlying causal facts (cf. Sober and Lewontin 1982). By stressing that evolutionary models are meant to be more than mere computational devices, realists resist the pluralists' inference from 'multiple representations' to 'no objective facts'.

Is there anything general that can said about multiple representation in relation to the levels of selection? Recent work by Kerr and Godfrey-Smith (2002) suggests that there is. They construct a simple model of evolution in a hierarchical setting, and show that the model's dynamics can be fully described by two sets of parameter values. The first set ascribes fitnesses only to the lower-level particles (in our terminology); the second set ascribes fitnesses to both the particles and the collectives. The latter is called a 'multi-level' parameterization, the former an 'individualist' parameterization.[19] Kerr and Godfrey-Smith demonstrate that the two parameterizations are mathematically equivalent—each set of parameter values can be derived from the other, so the overall evolutionary change can be expressed in terms of either. This is quite a general result, since their model is wholly abstract—it does not depend on what the particles and collectives are.

Kerr and Godfrey-Smith take their result to support pluralism, but of a modest variety. They argue that the multi-level and individualist perspectives are both valid ways of thinking about selection in a hierarchical setting, permitting different vantage points on a single evolutionary process. This is a weak form of pluralism because it says nothing about causality (as Godfrey-Smith and Kerr (2002) acknowledge). A realist could argue that in any given case, only one of the two parameterizations faithfully represents the underlying causal facts, despite their mathematical equivalence.[20] So Kerr and Godfrey-Smith's result does not in

---

[19] Kerr and Godfrey-Smith also describe the individualist parameterization as a 'contextual' parameterization, in recognition of the fact that individual fitnesses may depend on group context. However I prefer the label 'individualist', so as not to invite confusion with what was earlier called the contextual approach to MLS1.

[20] This is the essence of Sober and Wilson's (2002a) reply to Kerr and Godfrey-Smith's paper.

itself establish a strong pluralist thesis; but it does show, in a formally precise way, that strong pluralism is a coherent possibility. However, their result has one important limitation, discussed below.

How do multiple representations fit into the abstract framework developed in the previous chapters? One might think they do not fit in, given the framework's realist outlook. But in fact we have already encountered a simple case of multiple representation, which illustrates the essence of the Kerr–Godfrey-Smith result, in discussing the Price approach to MLS1.

Consider again a multi-level scenario, with particles nested in collectives, as in Figure 4.1 above. As we saw in Chapter 2, even though there are two hierarchical levels, it is possible to apply the single-level Price equation to the global population of particles. Ignoring particle transmission bias, this gives:

$$\overline{w}\Delta\overline{z} = \underset{\text{Global covariance}}{\text{Cov}(w_i, z_i)} \quad (4.1)$$

where $\Delta\overline{z}$ is the change in average particle character from one generation to another; w_i and z_i are the fitness and character of the ith particle respectively, and $\overline{w}$ is average fitness. Note that the index 'i' ranges over all the particles in the global population. Of course, this is tantamount to ignoring the collectives altogether: characters and fitnesses are attributed only to the particles, and the equation tells us only about the evolution of the particle character, in the global population of particles.

Recall that we then partitioned the global character-fitness covariance into within- and between-collective components, following Price (1972). This permitted us to expand equation (4.1) into:

$$\overline{w}\Delta\overline{z} = \underset{\text{Collective covariance}}{\text{Cov}(W_k, Z_k)} + \underset{\text{Average within-collective covariance}}{E_k(\text{Cov}_k(w_{jk}, z_{jk}))} \quad (4.2)$$

which is a multi-level equation, embodying the Price approach to MLS1. Note that in equation (4.2) the index 'i' has been replaced by the indices 'j' and 'k'. Recall that w_{jk} denotes the fitness of the jth particle within the kth collective, while W_k denotes the fitness of the kth collective itself, defined as the average fitness of the particles within it. So the equation deals with fitness at two hierarchical levels.

The fact that the total evolutionary change is correctly described by both equations (4.1) and (4.2) illustrates the basic Kerr and Godfrey-Smith point. Equation (4.1) corresponds to their 'individualist' parameterization—the particles are the sole fitness-bearing entities.[21] Equation (4.2) corresponds to their 'multi-level' parameterization—the particles and the collectives are both ascribed fitnesses, so two levels of selection are represented. The fact that equation (4.2) follows from (4.1) with no new biological assumptions mirrors the fact that Kerr and Godfrey-Smith's two parameterizations are inter-definable.

At this point a possible confusion should be forestalled. In Chapter 3 we argued that the Price approach to MLS1 is theoretically inferior to the alternative contextual approach, at least in most contexts. However, the difference between the Price and contextual approaches does not matter for the moment. For *both* of these approaches are multi-level, embodying the idea that the overall change depends on selection at two levels; they just disagree on how to chop up the change. So the Price and contextual approaches both correspond to Kerr and Godfrey-Smith's multi-level perspective.

This illustration of the Kerr and Godfrey-Smith result also brings out its limitation, namely that it *only* applies to multi-level selection of the MLS1 variety, in which the particles are the 'focal' entities and collective fitness is defined as fitness$_1$. When Kerr and Godfrey-Smith say that the evolutionary dynamics can be fully described using either parameterization, 'dynamics' refers to the change in mean *particle* character in the global population of particles—the quantity denoted above by $\Delta \bar{z}$. If 'dynamics' included the change in mean *collective* character, that is, the quantity affected by MLS2, then the claim about alternative parameterizations would not be true.

This limitation is unsurprising. As we have seen, in MLS2 the two levels of selection lead to incommensurable evolutionary changes, measured in different units and usually across different timescales. Given that collective fitness in MLS2 is defined as fitness$_2$, a quantity which bears no necessary relation to average particle fitness, it is impossible that the resulting evolutionary change could be expressed in terms of particle fitnesses alone. What permits this in MLS1 is the definitional link between particle and collective fitness. Therefore, multi-level selection of

---

[21] The only difference is that Kerr and Godfrey-Smith explicitly model the way that the w_i values are affected by collective membership, while equation (4.1) treats the w_i values as given.

the MLS2 variety *cannot* be re-described from Kerr and Godfrey-Smith's 'individualist' perspective.

This establishes a limit on the range of levels-of-selection problems to which pluralism might be applicable. Multi-level selection of the MLS1 sort always potentially invites a pluralist thesis, thanks to the equivalence of equations (4.1) and (4.2) above. This is not to say that pluralism is always, or indeed ever, correct in an MLS1 context, but it is at least a theoretical option. Multi-level selection of the MLS2 sort, by contrast, never invites a pluralist thesis, for it does not admit of multiple representation to start with.

Recall Sterelny and Griffiths's version of pluralism, discussed above. They argue for pluralism in cases where the collectives are loose aggregates of particles, but not in cases where the collectives are cohesive, physically bounded units. This idea has merit, but it needs qualifying. If particles and collectives are both focal units, that is, if our concern is with MLS2, then pluralism is *never* an option, as a matter of logic, whether or not the collectives are cohesive. Conversely, if particles are the sole focal units, that is, our concern is with MLS1, then pluralism is always potentially an option, since the evolutionary change can be described by a single-level or a multi-level equation.

This establishes the circumstances under which multiple representation arguments for pluralism are possible. But when are they correct? This is a trickier issue. It all depends on the status of the global character-fitness covariance Cov (w_i, z_i). In some cases this covariance will admit of a natural causal interpretation at the particle level, while in other cases it will be a statistical artefact. An example of the former is if a population of particles has been divided into collectives in a purely arbitrary way. A multi-level description of $\Delta \bar{z}$ will then be causally misleading; the correct explanation of why Cov (w_i, z_i) is non-zero need not mention the collectives. An example of the latter is where the collectives are cohesive, integrated entities. A single-level description of $\Delta \bar{z}$ will then be causally misleading, for the particles do not form a natural global population, subject to a uniform selection pressure.[22] In both of these cases pluralism (in its strong, causal version) would be wrong, despite the possibility of multiple representation.

This suggests that pluralism may be correct in intermediate cases, where neither the single-level nor the multi-level description can

---

[22] Giving a single-level description of what is intuitively a multi-level scenario, by averaging particle fitnesses across collectives, is an instance of what Sober and Wilson (1998) call the 'averaging fallacy'; see Chapter 6 Section 6.5.

immediately be dismissed as artefactual. What might such cases be like? It is important to see that the foregoing treatment of MLS1 does not answer this question, because it *assumes* a division of the particles into collectives. Given such a division, the analysis of cross-level by-products can be used to determine whether the global covariance, Cov (w_i, z_i), is due to particle or collective-level causal forces. This resolves the levels-of-selection issue, in a realistic rather than a pluralistic way, but *only* given the division in question. If there is a question mark over that division, or over the appropriateness of recognizing hierarchical structure at all, then matters are different. For then it becomes debatable whether we should even seek a multi-level description, rather than leaving the global covariance in its hierarchically un-decomposed form and attributing all the change to particle-level selection.

This means that multiple representation arguments for pluralism will only work in an MLS1 scenario where there is *also* indeterminacy of hierarchical structure—thus partially vindicating Sterelny and Griffiths. Once hierarchical structure is in place, and thus the propriety of seeking a multi-level description assumed, the level(s) of selection becomes a fully objective matter. A single-level description is still possible, by thinking in terms of a global population of particles, but it will be a statistical artefact. However, if the collectives are loose aggregates of particles, whose status as 'real' biological entities is in doubt, it may be unclear whether or not we should treat the global population of particles as hierarchically structured, and thus unclear whether a single- or multi-level description is preferable. Here, but only here, is there logical space for a 'no fact of the matter' claim about the level(s) of selection.

Our second and third sources of pluralism therefore coalesce. Indeterminacy of hierarchical structure, combined with the possibility of multiple representation that automatically arises on an MLS1 approach, are both necessary for pluralism to get a foothold; but neither is sufficient alone. This conclusion will prove useful when we examine the group selection controversy in Chapter 6.

Two final points deserve mention. First, there can be no *formal* answer to the question of when pluralism is and is not correct. Which types of evolutionary process can be multiply represented, and what the relations are between the alternative representations, are formal or quasi-formal questions. But the realism/pluralism issue depends on the underlying biological facts, not our representations of them, so cannot be resolved formally. Secondly, our analysis makes pluralism come out correct in a narrower range of cases than on Kerr and Godfrey-Smith's analysis.

This difference is the result of our adopting a stronger, explicitly causal notion of pluralism. Even in cases where our analysis says that there *is* an objective fact about the level(s) at which selection is acting, it may be true that viewing the selection process from more than one perspective is heuristically valuable, as Kerr and Godfrey-Smith urge.

4.4 REDUCTIONISM

The label 'reductionism' is often encountered in the levels-of-selection discussion, but its precise meaning can be difficult to pin down. For example, Williams's (1966) injunction that Darwinian explanations should always be sought at the lowest possible level is often called 'reductionistic', as is Dawkins's (1976) thesis that the gene is the 'real' unit of selection. In a different vein, Vrba (1989) and Sterelny (1996a) talk about selection at one level being 'reducible' to selection at another level; while Gould (2002) contrasts reductionism with what he calls 'hierarchical selection theory'. It is unclear, to say the least, that a single concept is at work here.

One standard view of reductionism in science construes it as an explanatory strategy that seeks to explain wholes in terms of their parts. Such a strategy is possible wherever there is part–whole structure, and has nothing essentially to do with Darwinism; it derives its rationale simply from the (presumed) fact of part–whole supervenience. For example, the idea known as 'methodological individualism' in the social sciences involves this sort of reductionism; it says that societal phenomena, for example, the higher divorce rate in some countries than others, should always be explained in terms of the behaviours of individual agents (Lukes 1968; Elster 1982). The contrast here is with holism, which tries to understand wholes 'at their own level', not via an understanding of their constituent parts.

Can this notion of reductionism be applied to the levels-of-selection question? I think it can. The difference between the MLS1 and MLS2 approaches to multi-level selection, outlined in Chapter 2, in effect corresponds to the reductionism/holism contrast. By taking the particles to be the sole focal entities, the MLS1 approach necessarily involves a 'bottom-up' mode of explanation (in so far as it aims to explain features of collectives). For example, MLS1 could explain the spread of a cooperative behaviour among individual ants, thus indirectly helping to explain complex social organization in ant colonies. So light is shed on a collective-level feature by explaining the evolution of the underlying particle characters on which the feature depends. By contrast, MLS2 treats the collectives as focal entities in their own right, bearing autonomously defined fitnesses; it thus contributes to an understanding of collective phenomena 'at their own level', rather than from the bottom up. An MLS2 explanation of the evolution of a collective character need assume nothing about how the character depends on underlying particle characters, so is inherently non-reductionist.

An interesting corollary of this is that the reductionism/holism issue, as it arises in relation to the levels of selection, has nothing to do with the issue of emergence. For as discussed previously, the MLS1/MLS2 contrast has nothing in particular to do with the distinction between aggregate and emergent characters (except in so far as fitness$_1$ and fitness$_2$ are themselves aggregate and emergent, respectively). MLS2 yields a 'holistic' mode of understanding, but the collective character on which it operates may nonetheless be aggregate. This is interesting because, historically, part of the motivation for holism has been the existence of emergent properties, and the apparent difficulty of explaining them in a reductionistic fashion.

Most models of group selection are of the MLS1 sort, as Damuth and Heisler (1988) note; their aim has usually been to explain the evolution of individually disadvantageous traits in group-structured populations. Such traits tend to be group beneficial, so explaining their evolution can contribute indirectly to an explanation of group-level functional organization. The underlying methodology here is clearly reductionistic. This partially vindicates the view of Gould (2002), who argues that the hierarchical view of evolution advocated by proponents of species selection constitutes a break with the reductionism of traditional neo-Darwinism. Gould is not alluding directly to the MLS1/MLS2 contrast, and does not say exactly which aspects of neo-Darwinism he regards as reductionistic. However, since species selection is of the MLS2 variety, and thus involves a holistic mode of explanation, there is some truth to what he says.

The choice between reductionism and holism is sometimes regarded as a matter of philosophical or scientific taste. But according to the foregoing analysis, MLS1 and MLS2 are causal processes which either do or do not occur; this appears to sit badly with the idea that it is up to us whether to adopt a reductionistic or holistic explanatory strategy. However, there is no real tension here. For any empirical enquiry must focus on some aspects of nature at the expense of others. So although it is an objective fact whether an MLS1 or an MLS2 process, or both, are at work in a given multi-level scenario, we may nonetheless choose to focus on one of these processes at the expense of the other. Conceivably, this choice could stem from an abstract preference for reductionism over holism or vice versa.

G.C. Williams's thesis that lower levels of selection are explanatorily preferable to higher levels is often regarded as reductionistic (Williams 1966, 1985); but this is not reductionism in the sense in which it contrasts with holism. Williams does not argue for a bottom-up approach to explanation in science, nor for the superiority of an MLS1 rather than an MLS2 approach to multi-level selection. Rather, he argues that natural selection is inherently more powerful at low hierarchical levels, and that empirically, most adaptations can be explained without appeal to higher levels of selection. These arguments have been much contested in the literature and partially retracted by Williams himself (cf. Wade 1978; Wright 1980; Williams 1992; Sober and Wilson 1998; Michod 1999). For the moment, the important point is that Williams's original thesis, and the research programme which it helped spawn, is based on an empirical belief about the types of selection process most likely to be found in nature, not on methodological considerations. For this reason it is not usefully described as reductionistic, in my view.

A quite different concept of reduction arises in the context of cross-level by-products. A cross-level by-product occurs where a character-fitness covariance at one level is a side effect of direct selection at another. In such cases, it can be natural to talk about selection at one level being 'reducible' to selection at another; conversely, the absence of a cross-level by-product can be described in terms of irreducibility. This mode of speaking was not used in Chapter 3, but is found in the literature. For example, in discussing species selection, Grantham (1995) and Sterelny (1996a) both ask whether differential speciation/extinction is 'reducible to organismic selection', as do Eldredge (1989) and Vrba (1989). In our terms, they are asking whether a character-fitness covariance at the species level is due to an organism→species by-product.

Importantly, 'reduction' in this context has nothing to do with reductionism qua explanatory strategy in science, and is not usefully opposed to holism. Indicative of this is the fact that cross-level by-products can run both up *and* down the biological hierarchy. Just as direct selection at the particle level can lead to a spurious character-fitness covariance at the collective level, so the converse is also possible. Downward-directed by-products could be described in terms of reducibility, but at the cost of a certain oddity; for it would mean that selection at one level was 'reducible' to selection at a *higher* level. (This is partly why the language of reduction was avoided in Chapter 3.) So reduction qua relation that may obtain between levels of selection, understood to mean the possibility of a cross-level by-product, has nothing to do with reductionism qua explanatory strategy for studying nature.

Yet another context in which 'reductionism' has been used is in the debate over genic selection and the gene's-eye view of evolution. The idea that all evolutionary change is at root a matter of genes being substituted for their alleles within populations is aptly called reductionistic. For intuitively, 'evolution' includes such phenomena as morphological and behavioural adaptation, the genesis of novel developmental programmes, speciation, and organic diversification, all of which seem only distantly related to gene frequency change. However, the precise sense in which the gene's-eye view is reductionistic, and the extent to which it is based on methodological rather than empirical considerations, are questions that can only be answered once the concept itself has been clarified. This is the task to which we turn next.

5

The Gene's-Eye View and its Discontents

INTRODUCTION

This chapter discusses the 'gene's-eye' view of evolution associated with Williams (1966), Dawkins (1976), and others. Section 5.1 outlines the origins of gene's-eye thinking. Section 5.2 argues for a distinction between the *process* of genic selection and the gene's-eye *perspective*, or viewpoint. Section 5.3 examines 'outlaw' genes, which benefit themselves at the expense of their host organism, leading to intra-genomic conflict. Section 5.4 revisits a theoretical issue from Chapter 3—the tension between the Price and contextual approaches to MLS1—and considers it in relation to genic selection. Section 5.5 looks at the well-known 'bookkeeping' objection, which says that the gene's-eye view merely records the outcome of evolution but says nothing about the underlying dynamics. Section 5.6 asks whether the gene's-eye view can accommodate epistasis and the context-sensitivity of gene action. Section 5.7 briefly revisits the topics of pluralism and reductionism.

5.1 THE ORIGINS OF GENE'S-EYE THINKING

Gene's-eye thinking can be traced back to the earliest days of evolutionary genetics. In his famous 1930 book, Fisher's attempt to synthesize Darwinism with Mendelian genetics led him to a novel conception of the evolutionary process. Instead of thinking in phenotype space, as Darwin had done, Fisher operated at the level of the underlying genes. He thought of natural selection as operating on a large population of genes, or gene pool, altering the pool's allelic composition over time. A separate selection coefficient, that is, fitness value, could be calculated for each allele, allowing organisms and their phenotypes to be bypassed. Evolutionary change, on this view, is simply gene frequency change, and natural selection is a force that leads fitter genes to be substituted for their alleles. Dawkins's concept of 'selfish genes' owes much to this Fisherian picture.

Fisher's picture is obviously an abstraction, as is the gene pool concept itself. In a review of Fisher's book, Wright (1930) questioned the utility of the abstraction on the grounds that 'genes favourable in one combination are...extremely likely to be unfavourable in another' (p. 84). Such epistatic effects mean that selection is not usefully thought of as operating separately at each locus, Wright argued. He allowed that individual genes could be ascribed selection coefficients, but regarded this as a computational convenience, not a reflection of the real causal forces at work. 'Selection relates to the organism as a whole and its environment and not to genes as such', Wright wrote (1930 p. 156). Contemporary opponents of gene's-eye thinking often make an argument strikingly similar to Wright's (cf. de Winter 1997).

The true utility of the gene's-eye approach only became apparent in the 1960s, thanks to Hamilton's work on the evolution of altruism. Darwin himself had realized that an organism which behaves altruistically will be at a selective disadvantage vis-à-vis its selfish counterparts, *ceteris paribus*. So it seems that altruism, and the genes which cause it, should be eliminated by natural selection.[1] Hamilton saw that the logic of this argument breaks down if altruistic actions are preferentially directed towards relatives. For relatives share genes, so there is a certain probability that the beneficiary of the altruistic act will *itself* carry the gene for altruism. To determine whether the altruism-causing gene will spread, we need to take into account not just the effect of the gene on the fitness of its bearer, but also on the fitness of the bearer's relatives.[2]

Hamilton realized that an even simpler way to determine whether a gene for altruism will spread is to forget about organismic fitness and think directly in genic terms, just as Fisher had done. He wrote: 'despite the principle of the "survival of the fittest", the ultimate criterion which determines whether a gene G will spread is not whether the behaviour [it causes] is to the benefit of the behaver, but whether it is to the

---

[1] In speaking of a 'gene' that causes altruism, we mean only that the gene increases the probability that its bearer will behave altruistically by some amount. This involves no presumption of genetic determinism, nor a downplaying of environmental effects on phenotype, as Dawkins (1982) rightly stresses.

[2] This is encapsulated in Hamilton's famous rule for the spread of altruism, $b/c > 1/r$, where c is the fitness loss incurred by the altruist, b the fitness gain enjoyed by the recipient, and r the coefficient of relationship between altruist and recipient. The proof of Hamilton's rule relies on certain non-trivial assumptions; see Michod (1982), Grafen (1985), Queller (1992a), or Frank (1998) for details.

benefit of the gene G' (1963 p. 7). So although altruism may seem anomalous from the organism's point of view, it makes perfect sense from the gene's point of view. Inducing its host organism to behave altruistically towards relatives is a 'strategy' that a gene can use to boost its representation in future generations.

Interestingly, Hamilton showed that altruism can in fact be understood from the organismic viewpoint too. Though behaving altruistically reduces an organism's *personal* fitness (by definition), it increases its *inclusive* fitness. An organism's inclusive fitness is defined as its personal fitness plus the sum of its weighted effects on the fitness of every other organism in the population, the weights being determined by the coefficient of relationship r. Given this definition, natural selection will favour those organisms with the highest inclusive fitness. So instead of thinking of genes trying to maximize the number of copies they leave, we can think of organisms trying to maximize their inclusive fitness. Most people find the gene's-eye approach more intuitive than the inclusive fitness approach, but mathematically they are equivalent (Michod 1982; Hamilton 1996; Frank 1998).

Gene's-eye thinking was developed further by Williams (1966) and Dawkins (1976), who argued that all organismic adaptations, not just pro-social behaviours, are ultimately for the benefit of the underlying genes. Dawkins also emphasized 'outlaw' genes, such as segregation-distorters and transposons, which spread despite their negative effects on the host organism's fitness (and thus on the fitness of all genes at unlinked loci in the same genome). Recent research has revealed outlaws, or 'selfish genetic elements' (SGEs), to be much more common than was originally thought; they constitute one of the strongest arguments for the utility of the gene's-eye view (Pomiankowski 1999; Hurst et al. 1996; Hurst and Werren 2001; Burt and Trivers 2006).

Dawkins offered another, quite different argument for treating the gene as the unit of selection, namely that genes are 'replicators'. Though entities at other hierarchical levels can reproduce, hence form parent–offspring lineages, the fidelity of reproduction is typically lower than that of DNA replication. This is especially true for sexually reproducing entities, where offspring contain a mixture of genetic material from two or more parents. Only genes have sufficient permanence to qualify as units of selection, Dawkins argued; organisms and their phenotypes are temporary manifestations. Despite the prominence Dawkins attached to this argument, arguably it confuses the unit of inheritance with the unit of selection. For as we saw in Chapter 1, replicators in

Dawkins's sense are not strictly needed for evolution by natural selection at all.

The influence of gene's-eye thinking has been enormous, particularly in behavioural ecology. Nonetheless, certain conceptual questions remain. Can all evolutionary change be understood from a gene's-eye viewpoint? Should the gene's-eye view be equated with genic selection, or are these concepts distinct? How does the gene's-eye view relate to the 'hierarchical' picture of evolution developed by multi-level selection theorists? These and other issues are explored below.

5.2 GENIC SELECTION AND THE GENE'S-EYE VIEW: PROCESS VERSUS PERSPECTIVE

Proponents of gene's-eye thinking have been guilty of a certain ambiguity. Sometimes they present their view as an empirical thesis about how evolution happened, sometimes as a heuristic perspective for thinking about evolution. Williams (1966) suggests the former interpretation, for he contrasts genic selection with group selection. Williams argues that if group-level adaptations turn out not to exist in nature, as he suspects, then 'we must conclude that group selection has not been important, and that only genic selection... need be recognized as the creative force in evolution' (p. 123–4). Dawkins's early work takes a similar line. Discussing Wynne-Edwards's theory that reproductive restraint in bird species evolved by group selection, Dawkins (1976) argues that a 'selfish gene theory' can explain the data better. The implication is that if Wynne-Edwards's theory were right, the selfish gene theory would be wrong and vice versa.

However, Dawkins (1982) adopts a different line, claiming not to be propounding 'a factual position... but rather a way of seeing facts' (p. vi). The selfish gene theory is not an empirical alternative to orthodox Darwinism, he claims; rather, it is an alternative *perspective* that is often heuristically valuable. We can think of evolution either in the orthodox way, in terms of selection between organisms (or other 'interactors'), or in the gene's-eye way, in terms of selection between genes. There is no empirical issue at stake—both are valid perspectives on a single set of facts.

The idea that the gene's-eye view is a heuristic perspective, not an empirical thesis, is closely bound up with the distinction between 'replicators' and 'interactors', discussed in Chapter 1. As we saw, Dawkins and Hull argue that replicators and interactors play complementary roles in the evolutionary process. Organisms and groups are interactors but genes are replicators; so to oppose genes to organisms, or to groups, as rival units of selection is to commit a category mistake. Organism and group-level selection are both *ways* by which genes can spread in a population, Dawkins argues (1984 p. 162). In a similar vein, Buss (1987) argues that there is no incompatibility between the gene's-eye view and multi-level selection; for any selection process, at whatever level, can also be viewed from a genic perspective.

This is a compelling analysis, but it raises certain questions. First, is it really true that a gene's-eye perspective is possible on *any* selection process? Why should this be so? This question is explored in Section 5.5. Secondly, if the gene's-eye view is simply a different way of thinking about orthodox Darwinism, what become of outlaw genes, which are *not* explicable in terms of advantage to the individual organism? The existence of outlaws formed part of Dawkins's original case, but they sit badly with the idea that the genic and orthodox approaches are equivalent.

This latter problem can be resolved by distinguishing sharply between selection processes that occur at the genic level, and a gene's-eye view on selection processes that occur at other levels. In cases of outlaws, the selection process itself is at the genic level—for there are fitness differences between the genes within the same organism.[3] In cases of organismic-, group-, or colony-level selection (for example) this is not so. But since selection at these higher levels typically leads to overall gene frequency change, it is possible to view the selection process from the gene's-eye perspective—even though the process itself does not take place at the genic level. So we must distinguish the *process* of genic selection, which is relatively infrequent, from the changes in gene frequency that are the *product* of selection at other levels, which are ubiquitous (or nearly so).[4]

The label 'genic selection' will therefore be reserved for selection between the genes within a single organism or genome, rather than for any selection process that leads to a gene frequency change. This understanding of 'genic selection' is increasingly used in the literature, for example, by Maynard Smith and Szathmáry (1995), Sober and

---

[3] Proponents of the replicator–interactor approach would express this by saying that where outlaws are concerned, the gene is both replicator and interactor at the same time; see Reeve and Keller (1999) for discussion of this move.

[4] The qualification 'nearly so' is necessary for reasons explained in Section 5.5.

Wilson (1998), Hurst and Werren (2001), Gould (2002), and others. It follows that genic selection must *not* be equated with the gene's-eye view; these are separate concepts.

Hurst (1996) observes that the expression 'selfish gene' has undergone a shift in meaning since Dawkins first introduced it. Originally it denoted *any* gene that spread by natural selection, irrespective of the selective mechanism; later it came to be used for outlaws or SGEs—which spread at the expense of other genes in the same genome. Obviously, on the former usage 'selfish genes' will be much more common than on the latter; for most genes that spread do so by cooperating, not competing, with the rest of the genome, that is, by selection on higher-level units. The ambiguity noted by Hurst corresponds precisely to the distinction between the process of genic selection, and a gene's-eye perspective on selection processes that occur at other levels.

Classical kin selection is an example of a process that does not occur at the genic level, but on which a gene's-eye perspective is nonetheless valuable. By inducing its host organism to behave altruistically towards relatives, a gene for altruism can spread in a population, as discussed. But the gene is not an outlaw—it does not harm the interests of other genes in the host organism. On the contrary, since donor and recipient have identical coefficient of relatedness at every locus in the genome, all the genes stand to gain equally from the altruistic behaviour.[5] So genic selection is not the force leading the altruistic gene to spread. However, it is still useful to adopt the gene's-eye view, and think of the altruistic behaviour as a strategy designed by the gene to boost its transmission. The alternative inclusive fitness approach to kin selection is much less intuitive, as noted above.

Buss (1987) attributes to M. Wade the remark: 'kin selection teaches not of the importance of the gene as a unit of selection, but of the family group as unit of selection above the level of the individual' (p. 184n). Wade's remark is insightful, for it highlights a confusion that many selfish gene theorists have fallen prey to, and ties in with the distinction between the process of genic selection and the gene's-eye perspective. (However Wade's own view—that kin selection is a type of group selection—is not universally accepted; others prefer to treat kin selection as type of individual selection, by regarding an individual's

[5] This assumes that donor and recipient are relatives in the ordinary sense. If they are not relatives but both happen to share the altruistic gene for some other reason then matters are more complicated, for the coefficient of relationship will then differ at different loci; see Okasha (2002) for discussion of this point.

relatives as part of its environment; see Chapter 6. But whichever of these views we favour, *neither* implies that kin selection involves selection at the genic level.)

The process/perspective distinction is crucial for assessing the objections to gene's-eye thinking. Numerous authors have pointed to biological phenomena, such as epistasis, heterosis, and epigenetic inheritance, which they claim cannot be accommodated by a gene's-eye approach (Wright 1980; Sober and Lewontin 1982; Sober 1984; Michod 1999; Jablonka and Lamb 1995; Avital and Jablonka 2000; Gould 2002). But it is not always clear whether they mean that the phenomena do not involve the *process* of genic selection, or cannot usefully be viewed from the gene's-eye *perspective*. These claims are quite different.

5.3 OUTLAWS AND GENETIC CONFLICTS

This section discusses outlaws and the genetic conflicts to which they lead. The focus is thus on the process of genic selection and its evolutionary consequences, rather than the gene's-eye perspective.

An outlaw, or SGE, is a gene that enjoys a transmission advantage over other genes in the same organism but does not increase the organism's fitness (Alexander and Borgia 1978; Dawkins 1982; Hurst et al. 1996; Hurst and Werren 2001). Most outlaws in fact reduce organismic fitness; this generates selection pressure at all unlinked loci for genes that suppress the outlaw's effects, leading to genetic conflict. Such conflicts are usually called 'intra-genomic', for they involve conflict between the different parts of a single genome. However, Maynard Smith and Szathmáry (1995) argue that some SGEs are better thought of as parasites or endo-symbionts, rather than as part of host genome, hence do not generate *intra*-genomic conflict, but rather conflict between the multiple genomes inhabiting a single organism.

Genic selection, as defined above, occurs when there are fitness differences between the genes within an organism. Given this definition, it follows that all outlaws spread by genic selection. But the converse is not necessarily true. A gene might boost the fitness of its host organism and *also* secure a transmission advantage over its allele. Such a gene would not be an outlaw, and would not induce selection for suppressors at unlinked loci; but its spread would still be due, in part, to genic selection. The effect of such a gene would be to benefit the whole genome, but to take a disproportionate share of the benefit for itself.

Defining genic selection this way requires that the notion of 'genic fitness' be appropriately understood. Consider a simple one-locus model with two alleles A and B, and thus three diploid genotypes AA, AB, and BB. Suppose that segregation is Mendelian—neither A nor B is a segregation-distorter. This means that no genic selection occurs—both alleles within an AB heterozygote have identical fitness, for they share the organism's gametic output equally. However, the *overall* fitness of the A and B alleles, averaged over all the genotypes, may well differ.[6] If so, one might be tempted to argue that there *are* fitness differences between the genes within an AB heterozygote, and thus that genic selection *is* occurring. But this would be a confusion. The crucial distinction is between the fitness of a token particle within a collective, and the fitness of that particle-type averaged across collectives. When we say that genic selection occurs if the genes within an organism differ in fitness, it is the former, not the latter, sense of 'genic fitness' that is pertinent.

This conception of genic selection sits well with the Price approach to multi-level selection. In Chapter 2, we saw how the Price approach can be applied to a one-locus model, by treating each diploid organism as a collective containing two particles. It then follows that organismic (or genotypic) selection occurs if there are fitness differences between organisms; while genic selection occurs if the alleles within the AB heterozygote differ in fitness, that is, if there is segregation distortion. The overall change in gene frequency depends on both levels of selection, as we saw.

This illustrates an important general point. By regarding an organism's genome as a 'group' of genes, and thus recognizing two potential levels of selection—within-group and between-group—one can apply the lessons of multi-level selection theory to the evolution of genomic organization. This approach has yielded substantial insights. For example, one lesson of multi-level theory is that the evolution of cooperative wholes requires suppression of competition among the parts (Frank 1995b; Michod 1999). This has been used to help explain why meiosis

---

[6] These overall fitness values, usually denoted w_A and w_B, are sometimes called 'marginal' allelic fitnesses; they determine which allele, if either, will increase in frequency. By contrast, the within-genotype allelic fitnesses determine whether or not there is a component of genic selection, i.e. whether or not segregation is Mendelian. See Kerr and Godfrey-Smith (2002), Okasha (2004a), and Section 5.5 for further discussion of the difference between the two types of genic or allelic fitness.

is usually fair, and why mitochondria are inherited uniparentally in the vast majority of eukaryotes (Haig and Grafen 1991; Hurst and Hamilton 1992). Both of these genomic features can be interpreted as adaptations for minimizing the damaging effects of lower-level selection on the integrity of the whole.

Many types of SGEs have been discovered, in both plants and animals. Ultimately, they all owe their existence to one of two related factors (Pomiankowski 1999; Maynard Smith and Szathmáry 1995). The first is that the different genetic entities within an organism, for example, autosomes, sex chromosomes, and organelles, do not always share the same mode of transmission. For example, mitochondrial genes are transmitted maternally while autosomal genes are transmitted biparentally. This creates a conflict of interest: mitochondrian genes favour a female-biased sex-ratio among the progeny of their host, while autosomal genes do not. The second is sexual reproduction. During meiosis, which precedes sexual fusion, only half the nuclear genes are passed to each gamete. So a gene which can increase its probability of surviving the meiotic cut will be selected, even if it reduces organismic fitness. This logic explains the behaviour of both segregation-distorters and transposons, which insert themselves at many chromosomal locations within the genome, thus boosting their chances of passing to a gamete.

Transposons are believed to be responsible for much of the non-coding DNA in the eukaryotic genome, a hypothesis first advanced by Doolittle and Sapienza (1980) and Orgel and Crick (1980). These authors argued that since transposition within the genome is sufficient to explain the prevalence of non-coding DNA, it is a mistake to think that such DNA *must* benefit the organism; it could simply be parasitic. This methodological point is clearly correct. In general, if selection at a given level can explain some feature, it is wrong to assume that that feature *must* benefit entities at another level; though cross-level synergism can occur. It is also possible that transposable elements were originally parasitic but later evolved more symbiotic relationships with their hosts (Pomiankowski 1999; Hoekstra 2003). In any case, the fact that transposons compose so large a fraction of the eukaryotic genome shows the evolutionary importance of selection at the genic level.

Further evidence of genic selection's importance comes from the discovery of segregation-distorters in numerous species. The best-known cases involve genes at two tightly linked autosomal loci, one producing a toxin, the other an antidote (Lyttle 1991; Maynard Smith and Szathmáry 1995). For example, in *Drosophila melanogaster*, an allele at the 'toxin' locus, denoted *Sd*, produces a product that inactivates any gametes that do *not* produce antidote. Whether a gamete produces antidote depends on which allele it has at a second locus; the Rsp^+ allele does not produce antidote, while the *Rsp* allele does. So if the two loci are tightly linked, then the *Sd*–*Rsp* pair constitutes an effective system for meiotic drive. Both alleles will achieve greater than fifty per cent representation in the successful gametes of their host organism, though they reduce organismic fitness.

This meiotic drive system, and similar systems in other species, depends crucially on the *Sd* and *Rsp* genes being in linkage disequilibrium; so recombination will tend to break it up. This is the basis of Haig and Grafen's (1991) suggestion that the function of recombination may be to prevent meiotic drive, and thus reduce the deleterious effects on organismic fitness. In effect, Haig and Grafen are suggesting that recombination evolved as a way of resolving the conflict between two levels of selection. Selection at the genic level led meiotic drive systems, consisting of a linked toxin–antidote pair, to evolve; because of the resulting negative effects on organismic fitness, selection at the organismic level led recombination to evolve, thus restoring fair meiosis.

A different type of conflict can occur between nuclear and cytoplasmic genes, arising because the latter are usually only transmitted maternally. Cytoplasmic outlaws exploit this fact to gain a transmission advantage for themselves at the expense of the nuclear genes. For example, in many angiosperm species, mitochondrial genes suppress male function—a phenomenon known as 'cytoplasmic male sterility'. The advantage to the mitochondrial gene is clear—it is transmitted in ovules but not pollen, so gains if the plant devotes more resources to making the former (Maynard Smith and Szathmáry 1995). Nuclear genes do not benefit from male sterility, since they *are* transmitted in pollen, so are selected to suppress the effects of the mitochondrial gene, that is, to restore male sexual function. Other forms of cytoplasmic outlawry include converting males into females, and male killing (Hurst et al. 1996).

Though uniparental inheritance of cytoplasmic genes leads to conflict over the sex ratio, many theorists believe that uniparental inheritance is itself an adaptation for eliminating another source of conflict (Hoekstra 1990; Hurst and Hamilton 1992). If mitochondria were biparentally inherited, then a mutant mitochondrian that abandoned normal cellular function for faster replication could gain access to the gametes and spread. Uniparental inheritance greatly reduces this problem, by ensuring that all the mitochondria in a cell are genetically similar, and making it easier for selection to weed out mutants. Thus the integrity of the whole is preserved by reducing the variance, hence opportunity for selection, among the parts. Ironically, uniparental inheritance eliminates one source of conflict—between variants in the cytoplasm—but creates another—between nuclear and cytoplasmic genes.

This brief survey of SGEs highlights a number of important points. The first is the importance of transmission asymmetry, which itself derives from sexual reproduction, in facilitating selection at the genic level. The second is that the 'interests' of different genetic elements converge to the extent that they are transmitted similarly, and diverge to the extent that they are transmitted dissimilarly. The third is the pervasive within-organism conflict to which outlaws give rise. Leigh (1977) observed that outlaws will typically be 'outnumbered' by the rest of the genome, so their effects will tend to be suppressed by the majority. This explains why organisms usually function as cohesive wholes despite the potential for conflict among their constituent genetic units. That is, genic selection may be relatively infrequent today precisely because organisms have evolved means to suppress it.

There is a minor irony lurking here. Outlaws are often regarded as the ultimate vindication of the gene's-eye approach, proof that the traditional organism-first paradigm cannot be sustained. There is something to this view. Although the origins of gene's-eye thinking lie in kin selection theory, which does *not* involve outlaws or genic selection, the gene's-eye perspective provides a framework into which outlaws fit easily. Indeed without that framework, it is hard to see how evolutionary theory could begin to make sense of outlaws. On the other hand, multi-level selection theory, which is sometimes regarded as the *antithesis* of gene's-eye thinking, is also crucial to understanding outlaws and genetic conflicts. The idea that a genome is a collective of genes with partially overlapping interests, that internal competition must be suppressed if the collective is to function as a whole, that selection at the collective level will favour such suppression, and that reducing the variation among the parts is one way to achieve it, are all themes from multi-level selection theory. Therefore, the gene's-eye and multi-level approaches are *both* needed to understand genomic organization; the two approaches are complementary, not antithetical.

5.4 PRICE'S EQUATION VERSUS CONTEXTUAL ANALYSIS REVISITED

Above we discussed how a one-locus population genetics model can be regarded as a multi-level system of the MLS1 type. The Price approach to MLS1 then applies neatly: genic selection operates on fitness differences between genes within organisms, while organismic selection operates on fitness differences between organisms. This implies that genic selection occurs if segregation in the AB heterozygote is distorted, that is, if there is meiotic drive. Since this corresponds to one standard conception of what 'genic selection' means, it is a point in favour of the Price approach.

However, recall the theoretical flaw with the Price approach discussed in Chapter 3: it fails to deal adequately with cross-level by-products. As we saw, contextual analysis addresses this flaw; it constitutes a rival to the Price approach to MLS1, for it partitions the overall change into two components in a different way. What happens when we apply the contextual approach to diploid population genetics, considered as a multi-level system?

Recall the contextual partition:

$$\overline{w}\Delta \overline{z} = \beta_2 \, \text{Var}(Z) + \beta_1 \, \text{Var}(z)$$

with labels: Collective-level selection ($\beta_2 \text{Var}(Z)$), Particle-level selection ($\beta_1 \text{Var}(z)$).

where z is particle character, Z is collective character (defined as mean particle character), w is particle fitness, and β_1 and β_2 are the partial regressions of w on z and Z respectively. It is perfectly possible to apply this partition, rather than the Price partition, to the diploid population genetics model. But interestingly, doing so produces extremely counter-intuitive results.

To see why, consider a situation analogous to the case of pure 'soft selection' discussed in Chapter 3. The three diploid genotypes, AA, AB, and BB, have identical fitnesses, that is, $w_{AA} = w_{AB} = w_{BB}$. But there are 'organismic effects' on genic fitness—an A allele in an AA homozygote has lower fitness than an A allele in an AB heterozygote. This means that segregation in the heterozygote is distorted in favour of A. Intuitively, all the selection is at the genic level in this example—for the organisms themselves do not differ in fitness. However, the contextual approach will detect a component of organismic selection, for differences in organismic character will help predict differences in genic fitness, controlling for genic character.[7] So β_2 in the equation above will be non-zero, indicating selection at the organismic level.

This is not the only unpalatable consequence of the contextual approach as applied to diploid population genetics. Consider the following example. Genotypic fitnesses are $w_{AA} = 16$, $w_{AB} = 12$, and $w_{BB} = 8$. Segregation is distorted in favour of the A allele in the ratio 8:4, that is, of the 12 gametes that an AB organism contributes to the next generation, 8 are A, and 4 are B. Given this fitness scheme, contextual analysis implies that *all* the selection is at the genic level. For the fitness of a gene is independent of its organismic context—an A gene has a fitness of 8, irrespective of which organism it is in, and a B gene has a fitness of 4, irrespective of which organism it is in.[8] So β_2 equals zero, implying that genic selection is the only force in operation. This is deeply implausible, and not something that any evolutionist would want to say.

In short, if we wish to treat diploid population genetics as a multi-level system, the Price approach seems clearly preferable to the contextual approach. This is interesting for two reasons. First, it shows that despite the theoretical argument in favour of the contextual approach, in some cases it gets the answer 'wrong' while the Price approach gets it 'right'. Secondly, it shows that multi-level systems that are formally isomorphic, but have different biological interpretations, can elicit from us very different intuitions about the level(s) at which selection is acting. This point is worth expanding on.

Recall the case that motivated contextual analysis originally: the particles are organisms, the collectives are groups, and there are no 'group effects' on organismic fitness, that is, the fitness of an individual organism depends only on its own phenotype. Biologists are unanimous that there is no group selection in this scenario—all the selection is at the lower level. But the case described in the paragraph before last, where the particles are genes, the collectives are diploid organisms, and there are no 'organismic effects' on genic fitness, is formally identical.

---

[7] To see this, note that if you are trying to predict the fitness of a randomly picked gene from the population and you already know whether it is A or B, additional information about the genotype of its host organism *does* help you make your prediction.

[8] The crucial feature of this example is that segregation in the AB heterozygote is distorted in favour of the A allele in the ratio w_{AA}/w_{BB}. Wherever this condition is satisfied, then a gene's fitness will be independent of its organismic context.

However, in this case it seems crazy to assert that all the selection is at the genic level and none at the organismic level. The two cases are formally isomorphic, but they elicit very different intuitions about the levels of selection.

Why is this? The answer, I think, is that the formal isomorphism belies a biologically important difference. In the group selection case, the question we are critically interested in is whether there are 'group effects' on organismic fitness. In the diploid population genetics case, we are not especially interested in whether there are 'organismic effects' on genic fitness. The situation described three paragraphs back, where $w_{AA} = 16$, $w_{AB} = 12$, $w_{BB} = 8$, and segregation is distorted in the ratio 8:4 in favour of A, is of no theoretical significance at all. This is because the explanation of *why* the fitness of an A gene is the same, whatever its organismic context, involves two quite disparate circumstances: the fact that segregation is distorted in a certain very specific way, and the fact that genotypic fitnesses are as they are. By contrast, where the fitness of an organism depends on its own phenotype alone, irrespective of group context, this *is* theoretically significant—it signals the absence of group effects on fitness. So although the fitness structures in the two cases are isomorphic, biologically they are quite unlike.

The disanalogy can be seen in another way. In the group selection case, fitnesses are possessed in the first instance by individual organisms. Group fitness is derivative—a group only has a fitness in virtue of the fitnesses of its constituent organisms. (Recall that we dealing with MLS1). In the diploid population genetics case it is the other way round. Fitnesses are again possessed in the first instance by individual organisms—but they are now the collectives, not the particles. The fitness of a gene within an organism, defined as the number of copies of the gene in the organism's successful gametes, is derivative—a gene only has a fitness in virtue of its host organism's fitness. So the biological explanation of why the fitnesses values are as they are is quite different in the two cases.

What moral should we draw? One might conclude that diploid population genetics should not be treated as a multi-level system at all. After all, diploid organisms are not *really* groups of two genes; to regard them as such is an idealization. But this conclusion is over hasty. The basic idea of treating an organism's genome as a group of genes and then applying multi-level selection theory has considerable explanatory power, as discussed. And the idea that meiotic drive constitutes selection at the sub-organismic level is widely accepted. So although the gene–organism relation is in some respects unlike the organism–group relation, there seems nothing *in principle* wrong with treating diploid population genetics as a multi-level system.

I think the correct moral is twofold. First, since the Price approach sometimes works better than the contextual approach, theoretical arguments for the latter notwithstanding, there cannot be a fully *general* solution to the problem of causally decomposing the total evolutionary change in an MLS1 scenario. Secondly, the fact that two fitness structures can be formally isomorphic, yet generate different intuitions about the levels of selection, shows that the biological interpretation of the fitness structure is also crucial (cf. van der Steen and van den Berg 1999). This indicates a limit on the extent to which the levels-of-selection question can be addressed in purely abstract terms.

Finally, recall the discussion of the Wimsatt/Lloyd 'additivity criterion' in Chapter 4. There we showed that the relevance of additivity is different, depending on whether we favour the Price or the contextual approach to MLS1. On the Price approach, additivity of variance is wholly irrelevant to determining the level(s) of selection, but on the contextual approach it is partly relevant. We treated this as a partial vindication of Wimsatt and Lloyd, given the theoretical superiority of the contextual approach. But the foregoing results complicate matters, given that the contextual approach yields the 'wrong' answer when applied to diploid population genetics.

In a one-locus model, perfectly additive variance in collective fitness means the absence of any dominance or heterosis, that is, genotype fitness is a linear function of allelic 'dosage', so $(w_{AA} - w_{AB}) = (w_{AB} - w_{BB})$. Certain theorists, notably Wright (1980), have argued that the difference between 'genic' and 'organismic' selection does indeed depend on whether the variance in genotype fitness is additive. But if we agree that the Price approach is the correct way of applying multi-level theory to diploid population genetics, Wright's conclusion must be rejected. Whether selection is at the genic or organismic level depends on whether the fitness differences are within or between organisms, or both; this has nothing to do with the additivity of variance.[9]

---

[9] This explains why our partial defence of the additivity criterion in Chapter 3 is compatible with the criticisms of Wimsatt and Lloyd made by Sarkar (1994) and Godfrey-Smith (1992). For these authors' criticisms are directed against the additivity criterion as applied to diploid population genetics; and in this context the criterion does not work, for the reasons given in the text.

5.5 BOOKKEEPING AND CAUSALITY

I turn now to one of the central objections to the gene's-eye view: that it or ignores, or obscures, the causal structure of selection processes. Construing evolution as the substitution of fitter genes for their alleles is all very well, the objection goes, but it tells us nothing about *why* the substitutions have occurred. Thus gene's-eye theorists are accused of 'confusing bookkeeping with causality'. Versions of this objection have been made by Wright (1930), Mayr (1963), Sober and Lewontin (1982), Sober (1984), Brandon (1990), Gould (2001, 2002), and others.

Proponents of this argument usually allow that the *outcome* of selection can be described in genic terms. Thus for example, Gould (2001) accepts that gene frequencies provide the best way of 'keeping the books' of evolution, since selection at all levels does eventuate in gene frequency change, he thinks. So the gene's-eye perspective is always available, he argues, but rarely useful, for it omits causal information.

There are two separate issues here. First, is it true that the outcome of *any* selection process can be described in genic terms? Is Gould right to concede this point to the gene's-eye theorists? Secondly, in cases where the gene's-eye perspective is available, what determines whether it is useful? I tackle these questions in turn.

5.5.1 The Limits of Genic Accounting

It is clear that natural selection, at various levels, does produce gene frequency change. For example, genic selection leads to the spread of genes which boost genic fitness, for example, segregation-distorters; organismic selection to the spread of genes which boost organismic fitness; and group selection to the spread of genes which boost group fitness, for example, genes for pro-social behaviour. So a gene's-eye perspective on the evolutionary changes produced by these modes of selection is clearly possible.

However, the gene's-eye perspective will not be universally available, for two reasons. The first concerns non-genetic inheritance. For natural selection to produce an evolutionary response, offspring must tend to resemble their parents with respect to the character in question. Shared genes are one cause of parent–offspring resemblance, but not the only cause. Cultural transmission, behavioural imprinting, and other types of epigenetic inheritance can also generate parent–offspring resemblance (cf. Boyd and Richerson 1985; Avital and Jablonka 2000). Clearly, where evolutionary change is mediated by non-genetic inheritance mechanisms, gene frequencies may be unchanged, so the gene's-eye perspective will be unavailable. Evolution of this sort cannot be understood as the result of 'selfish' genes striving for advantage over their alleles (cf. Jablonka 2001).

Proponents of gene's-eye thinking often argue that these phenomena are unimportant compared with ordinary gene-based evolution (Dawkins 2004; Cronin 2005). The issue here is basically empirical, for no one denies that non-genetic inheritance *can* lead to adaptive evolution; the question is how often it has done so. This question cannot be tackled properly here; but two recent arguments for the evolutionary importance of non-genetic inheritance deserve mention, for they bear directly on the levels-of-selection issue.

A number of theorists have suggested that human cultural inheritance provides a particularly favourable context for group selection to work (Boyd and Richerson 1985, 2005; Wilson 2002; Heinrich et al. 2003). One traditional objection to group selection is that migration will tend to eliminate the between-group genetic variation, reducing the potential for group selection by making the groups too alike. But an analogous objection does not necessarily apply to cultural variation. Migrants are often forced to convert to the culture of the group they are entering, so even substantial migration need not dilute the cultural differences between groups. Once cultural inheritance is taken into account, the hypothesis that humans may be a heavily group-selected species, long regarded with suspicion by evolutionists, demands reconsideration.

Jablonka (1994) takes this line of thought a step further. She argues that cultural inheritance has led to the evolution of a new biological individual: the human society. Interestingly, she suggests that other biological individuals, including multicelled organisms, also evolved with the help of non-genetic inheritance. When cells divide, daughter cells inherit not just the DNA sequences of the genes in the parent cell but also the genes' functional states; this is the basis of cellular differentiation in modern organisms. Jablonka argues that though epigenetic cellular inheritance first evolved in single-celled organisms, it facilitated the transition to multicellularity. For epigenetic inheritance tends to make groups of related cells phenotypically homogenous, and distinct from other groups, thus allowing selection between cell-groups to dominate selection within cell-groups. This in turn promoted the evolution of cohesive multicelled creatures, just as cultural inheritance promotes the evolution of cohesive human societies.

If these arguments are correct, they suggest that non-genetic inheritance may play a role in evolutionary transitions, by providing conditions under which higher-level selection can operate effectively. If so, then a purely gene's-eye approach to the evolutionary transitions will not be fully adequate. However, the jury is still out on both arguments.[10]

The second reason why a gene's-eye perspective is not always available is quite different. As Sober (1984) notes, gene frequencies are normally computed *per organism*. If an organism bearing a mutant gene gets fatter, we do not say that the gene has increased in frequency, even though there are now more cells bearing the gene (p. 30). This is not the only possible counting convention, and one might question its motivation. Why privilege organisms this way? Why not count on a per cell basis, or a per colony basis, for example? Part of the answer is that what we usually want to explain, in evolutionary theorizing, is the incidence of organismic phenotypes, so organisms are the 'focal' units. When phenotype frequencies change, gene frequencies, computed per organism, will usually change too.

However, organisms are not *always* the focal units, as discussed in Chapter 2; so if we stick to the 'per organism' counting convention, there are bound to be evolutionary changes that do not involve changes in gene frequency, hence cannot be viewed from a gene's-eye perspective. (Arnold and Fristrup (1982) illustrate this with a hypothetical example of species selection, in which there is differential species extinction but no associated changes in (per organism) gene frequency.) However, a gene's-eye theorist could reply that where organisms are *not* the focal units, the standard counting convention needs altering. Perhaps the gene's-eye perspective will always be available, so long as genes are counted 'per focal unit'.

This suggestion is logical, but problematic. The standard 'per organism' counting convention relies on the fact that a typical organism begins life as a single cell, containing either one or two copies of the genetic material. So genes are counted at the zygotic stage. However, not all entities in the biological hierarchy have life cycles with single-celled bottlenecks. Some entities reproduce by budding, fission, or propagule

---

[10] See Heinrich et al. (2003) for critical discussion of cultural group selection, and Michod (1998) for discussion of Jablonka's argument about the role of epigenetic cellular inheritance in the evolution of multicellularity.

emission; still others are formed by aggregation of a large number of smaller entities. If our focal units have life cycles of this sort, it is unclear what it would *mean* to count genes 'per focal unit'.

To see the problem, recall the discussion of clonal plants from Chapter 2. We noted that biologists disagree about whether to treat genets or ramets as the bearers of Darwinian fitness, that is, as the focal units. Genet-based theorists argue that clonal production of ramets constitutes growth rather than reproduction; ramet-based theorists insist that ramets are genuine evolutionary individuals, with life cycles of their own, despite their asexual origin. If genets are our focal units, then the genic accounting is straightforward. Since each genet starts life as a single cell, we can count genes on a per zygote basis in the usual way. Matters are different with ramets. They are usually produced vegetatively, so begin life as multicellular units (meristems), which are often chimeric. So it is not possible to count genes 'per meristem', for the cells that compose the meristem may have different genotypes. If ramets are our focal units, there is no way of doing the genic accounting, nor therefore of adopting the gene's-eye perspective on the evolutionary change in question.

This problem is not unique to ramets; it arises wherever the focal units do not have life cycles with bottlenecks. Take social insect colonies, for example. Some colonies are founded by a single, once-mated queen; the life cycle of these colonies passes through a relatively narrow bottleneck, though not a single-celled one. But other colonies are founded by multiple or multiply-mated queens, and are thus composed of several genetically distinct lineages. If our focal unit was a colony of the latter sort, there would be no way of counting genes on a per colony, rather than a per insect, basis. Unsurprisingly, most work on social insect evolution treats the individual insects, not the colonies, as the focal units—even when the aim is to explain colony-level properties.

Maynard Smith and Szathmáry (1995) argue that 'a gene-centred approach' is necessary to understand the major evolutionary transitions. They write: 'there is in fact one feature of the transitions . . . that leads to this conclusion. At some point in the life cycle, there is only one copy, or very few copies, of the genetic material' (p. 8). I agree that this feature leads naturally to the gene-centred approach; without it, no clear meaning attaches to 'gene frequency'. But it is clearly *not* true that all biological entities have life cycles with bottlenecks. One of Maynard Smith and Szathmáry's own examples of a major transition is from primate to human societies (p. 6). But the 'life cycle' of a human society does *not* pass through a stage at which only one or a few copies of the genetic material is present. It would make no sense to try to count genes on a 'per society', rather than a per human, basis.

To summarize, there are two reasons why the gene's-eye perspective will not always be available. First, some evolutionary changes are mediated by non-genetic inheritance, so do not involve gene frequency change at all. Secondly, gene frequencies are normally computed per organism; but this mode of accounting presumes that organisms are the focal units. Where the focal units are entities at other hierarchical levels, it may not be possible to count genes 'per focal unit'; it depends on their mode of reproduction. This delimits another class of evolutionary changes that cannot be viewed from the gene's-eye perspective.

5.5.2 Sober and Lewontin's Heterosis Argument

The previous section asked whether it is always *possible* to view evolution from the gene's-eye perspective. This section asks whether it is always *desirable* to do so, when possible. Those who answer 'no', to recall, argue that the gene's-eye view misleads about the causal structure of selection processes. One of the clearest versions of this argument was made by Sober and Lewontin (1982), in relation to the phenomenon of heterozygote superiority, or heterosis.

Sober and Lewontin consider a one-locus model with two alleles A and B, whose initial frequencies are p and q respectively. There are three genotypes in the population, AA, AB, and BB, with pre-selection frequencies in Hardy-Weinberg equilibrium. Genotypic fitnesses, assumed constant, are w_{AA}, w_{AB}, and w_{BB}.

	AA	AB	BB
Initial frequency	p^2	$2pq$	q^2
Fitness	w_{AA}	w_{AB}	w_{BB}

To calculate post-selection frequencies, we multiply initial frequency by fitness for each genotype, then normalize by dividing by mean population fitness $\overline{w}$ in the usual way; where $\overline{w} = (p^2 w_{AA} + 2pq w_{AB} + q^2 w_{BB})$. This gives:

	AA	AB	BB
Post-selection frequency	$(p^2 w_{AA})/\overline{w}$	$(2pq w_{AB})/\overline{w}$	$(q^2 w_{BB})/\overline{w}$

New allelic frequencies are $p' = (p^2 w_{AA} + pq w_{AB})/\bar{w}$; $q' = (q^2 w_{BB} + pq w_{AB})/\bar{w}$. If p' and q' are different from p and q, then evolution by natural selection has occurred: one allele has increased in frequency at the expense of the other.

In this simple model, fitnesses are ascribed to diploid genotypes, not individual genes. But as Sober and Lewontin note, it is straightforward to calculate overall genic fitness coefficients w_A and w_B which suffice to predict the system's evolution. To do this, we simply average the fitness of the A allele across all the diploid genotypes in which it occurs, weighting by its relative frequency within each genotype (0, $1/2$, or 1). Presuming meiosis is fair, this gives:

$$w_A = (w_{AA} p^2 + w_{AB}(1/2)2pq)/(p^2 + (1/2)2pq) = p w_{AA} + q w_{AB}$$
$$w_B = (w_{BB} q^2 + w_{AB}(1/2)2pq)/(q^2 + (1/2)2pq) = q w_{BB} + p w_{AB}$$

These genic selection coefficients now predict the evolution of the system. If $w_A > w_B$ then the A allele will spread; if $w_B > w_A$ then the B allele will spread; if $w_A = w_B$ then the system is in allelic equilibrium. So even though fitnesses were initially ascribed to diploid organisms, the resulting evolutionary change can be viewed from the gene's-eye perspective.

Importantly, w_A and w_B are the *overall* (or 'marginal') genic fitnesses, found by averaging the fitnesses of the A and B alleles across genotypes. In Section 5.3 we used 'genic fitness' to mean the fitness of a gene *within* a genotype, in order to discuss meiotic drive; in that sense of 'genic fitness', there are no differences in genic fitness in Sober and Lewontin's example, for meiosis is stipulated to be fair.

Gene's-eye theorists often emphasize the significance of coefficients such as w_A and w_B. Since they contain all the information needed to predict the system's evolution, they allow us to view the evolutionary change as the result of competition between the two alleles. For example, G. C. Williams (1966) wrote: 'one allele can always be regarded as having a certain selection coefficient relative to another at the same locus at any given point in time... Adaptation can thus be attributed to the effect of selection acting independently at each locus' (p. 112).

Sober and Lewontin argue that in some cases, averaging over genotypes and attributing evolutionary change to differences in genic fitness is unexceptionable. However, in cases of heterosis, that is, where $w_{AB} > w_{AA}$ and $w_{AB} > w_{BB}$, the averaging strategy obscures the causal structure of the selection process. In such cases, w_A and w_B can still be calculated, but they are statistical artefacts, Sober and Lewontin argue. Selection is really acting on the diploid organisms, not the individual genes.

What is so special about heterosis? The crucial feature, according to Sober and Lewontin, is that the effect of each allele on organismic fitness is context-dependent. In some genotypic contexts, the A allele *raises* an organism's fitness, while in other genotypic contexts it *lowers* an organism's fitness. The A allele therefore does not have a 'uniform causal role', and so cannot itself be subject to selection; the same is true of the B allele. 'If we wish to talk about selection for a single gene', Sober and Lewontin write, 'then there must be such a thing as the causal upshot of possessing that gene' (1982 p. 122). This is the condition that is violated in the heterosis example, and is why the averaging strategy misleads about the level of selection.

Sober and Lewontin's argument implies that it *is* permissible to regard the gene as unit of selection so long as each allele *does* have a 'uniform causal role'. For a gene to have a uniform causal role, the presence of the gene must raise (lower) genotypic fitness in at least one genotypic context, and not lower (raise) it in any context (Sober 1984). This means that genotype fitness must be a monotonic function of allelic dosage, that is, substituting an A for a B allele must always affect genotypic fitness in the same direction. With heterosis, genotypic fitness depends non-monotonically on allelic dosage, as shown in Figure 5.1.

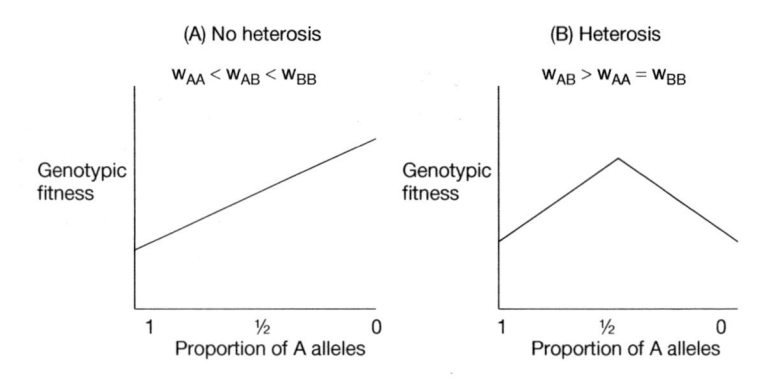

(A) No heterosis
$w_{AA} < w_{AB} < w_{BB}$

(B) Heterosis
$w_{AB} > w_{AA} = w_{BB}$

Figure 5.1. Genotypic fitness as a function of allelic dosage

Sober and Lewontin's argument has been extensively discussed in the literature. Some commentators have endorsed the argument; others have rejected it, claiming that heterosis involves no more than the standard relativity of fitness to environment—genetic environment in this case. Interestingly, Maynard Smith (1987a) says that he was initially convinced by the heterosis argument but later changed his mind.

The most detailed response to Sober and Lewontin comes from Sterelny and Kitcher (1988) and Sterelny and Griffiths (1999). These authors make two points, one correct, the other incorrect. Their correct point is that, from a gene's point of view, the other allele in the diploid genotype is part of its environment; and the fitness of any biological unit is environment-relative. However, Sterelny and Kitcher (1988) also claim a link between heterosis and frequency-dependence, which is a mistake. They say that with heterosis, the fitness of each gene becomes a function of its frequency in the population (p. 343). They then analogize heterosis to the hawk–dove game of ESS theory, the classic example of a frequency-dependent fitness structure. Since the hawk–dove game can be understood from an 'individualist' perspective, so heterosis can be understood from a 'genic' perspective, they argue.

This argument is flawed for a simple reason, namely that the overall genic fitnesses are *always* frequency-dependent, heterosis or no. This can be seen simply by inspecting the above expressions for w_A and w_B. Both w_A and w_B are functions of p and q (except in the degenerate case where $w_{AA} = w_{AB} = w_{BB}$). So Sterelny and Kitcher are wrong to say that heterosis leads the genic fitnesses to become frequency-dependent; they are frequency-dependent anyway.[11]

This brings us to the real problem with Sober and Lewontin's argument, which is that it focuses on the wrong phenomenon. They argue that a gene's-eye perspective obscures the true level of selection. A stronger argument for this conclusion would focus on meiotic drive, not heterosis. If we presume that meiosis is fair, as Sober and Lewontin's model does, all the selection must be at the organismic level—for all the fitness differences are between, rather than within, organisms (as Sober and Wilson (1994) acknowledge). This is true *whether or not there is heterosis*. Meiotic drive, however, introduces another level of

---

[11] Strictly speaking, this conclusion needs qualifying. The above expressions for w_A and w_B assume that meiosis is fair. If meiosis is not fair, it is theoretically possible for the overall genic fitnesses to be independent of frequency. Recall the example from Section 5.3, where genotypic fitnesses are $w_{AA} = 16$, $w_{AB} = 12$, and $w_{BB} = 8$, and segregation is distorted in the ratio 8:4 in favour of A. In this case, w_A and w_B are 8 and 4 respectively, whatever the allelic frequencies. But this is a special case, arising from a specific pattern of segregation-distortion and jury-rigged genotype fitnesses. With fair meiosis, genic fitnesses are always frequency-dependent.

selection, and thus the potential for a more serious obscuring of the causal structure.

What Sober and Lewontin should have argued, therefore, is this. The values of w_A and w_B determine which allele is fitter overall, and so which will spread. But this tells us nothing about the level at which selection acts. Suppose $w_A > w_B$. This might be because of selection acting at the organismic level, for example, if $w_{AA} > w_{AB} > w_{BB}$. Or it might be because of selection at the genic level, for example, if the A allele is a segregation-distorter. Or it might be due to selection at both levels. Merely focusing on the overall genic fitnesses cannot discriminate these three possibilities, though it correctly predicts the gene frequency change. So the gene's-eye perspective is predictively adequate, but explanatorily inadequate.

This conclusion is identical to Sober and Lewontin's but arrived at in a different way. Sober and Lewontin consider a diploid model with fair meiosis and contrast two cases—one with heterosis, one without. But this is not the right contrast, for in *both* these cases, selection acts only at the organismic level. The relevant contrast is between a case with fair meiosis and one with meiotic drive. For these cases involve a difference in the level of selection—and yet both can lead to the same values for w_A and w_B. So Sober and Lewontin's conclusion is quite right, though their argument for that conclusion is flawed.

Reformulating Sober and Lewontin's argument this way shows it to be a special case of a more general point, discussed previously. In a multi-level scenario of the MLS1 type, the total evolutionary change depends on the *global* character-fitness covariance, in the overall population of particles. This global covariance can be decomposed into collective-level and particle-level components, via the Price equation (or via contextual analysis). So merely computing the global covariance, though sufficient to predict the overall change, does not tell us the level at which selection acts. In the one-locus diallelic model, the fact that the observed values of w_A and w_B can result from genic selection, organismic selection, or a combination, is an instance of this general lesson from multi-level theory.

5.6 CONTEXT-DEPENDENCE AND THE GENE'S-EYE VIEW

The previous section criticized Sober and Lewontin's heterosis argument. However, it may be possible to read their argument in a different way.

Perhaps we can read Sober and Lewontin as supplying a criterion, not for whether selection occurs at the genic level, but for whether the gene's-eye perspective is heuristically useful. So construed, the argument would say that the gene's-eye perspective is only useful if genotypic fitness is a monotonic function of allelic dosage, that is, if there is no heterosis. Otherwise, the gene's-eye perspective is not useful.

This is actually quite plausible. Thinking in terms of a gene's 'interests', as Dawkins and others do, makes good sense where a gene has a relatively context-independent effect on its host organism. The gene's phenotypic effect can then be thought of as a 'strategy' it has devised to promote its transmission. But as critics have noted, if a gene's phenotypic effect is highly dependent on context, including genetic context, it is misleading to think this way, since the gene's chances of transmission are not in its own hands—they depend on extraneous factors. And where monotonicity is violated, we have an extreme of context-dependence—the *type* of effect a gene has on host fitness depends on which allele it happens to be paired with. So the utility of adopting the gene's-eye view is somewhat compromised by heterosis.

Note that this argument has nothing to do with causality. The argument is *not* that if there is heterosis, the gene's-eye perspective will mislead about the causal structure of the selection process. As we have seen, the causal structure of the selection process, that is, whether it occurs at the genic or the organismic level, has nothing to do with heterosis. Rather, the argument is that if there is heterosis, the utility of adopting the gene's-eye perspective is undermined.

A similar type of argument could be made in relation to additivity, rather than monotonicity. In Section 5.4 we saw that whether the variance in genotypic fitness is additive or not implies nothing about whether selection is acting at the genic level. However, additivity may be relevant to whether a gene's-eye perspective is heuristically useful. Such a suggestion has sometimes been made. Thus for example, Gould (1999) argues that the gene's-eye view would only make sense 'if genes built organisms entirely in an additive fashion with no nonlinear interaction among genes at all' (p. 213). Wimsatt (1980) argued similarly, as did Wright (1930) in response to Fisher.

The idea that the heuristic utility of the gene's-eye view requires additivity is related to the idea that it requires monotonicity, since the former is a special case of the latter. In a one-locus diploid model, requiring additivity would lead us to adopt the gene's-eye view only if there were no dominance at all, which is obviously a rarer circumstance than the absence of heterosis. Generalized to the multi-locus case, requiring additivity would lead us to adopt the gene's-eye view only if there were no epistatic interactions between genes at different loci, which is a yet rarer circumstance.

How should we decide whether monotonicity or additivity, if either, is the relevant requirement? I see no clear way of answering this question. A hard and fast criterion for when the gene's-eye perspective is heuristically useful is unlikely to be found, since whether one finds a perspective useful is partly a subjective matter. Nonetheless, it is true that Dawkins's vision of evolution — genes manipulating the world around them for their own benefit — applies most naturally when genes have relatively context-independent effects on phenotype and fitness. The metaphors of manipulation and agency become strained when a gene's effects are heavily context-dependent. Non-additive interaction between genes, whether at the same or different loci, generates context-dependence of this sort; and heterosis is simply an extreme of non-additivity.

It is worth briefly noting how gene's-eye theorists have handled non-additivity and context-dependence. One response, due to Williams (1966), notes that we can always consider the effect on organismic fitness of substituting one gene for its allele against a given genetic background, however much epistasis there is. This point is correct, but of dubious relevance. For in a sexually reproducing species, genes recombine every generation, so natural selection cannot 'try out' two alleles against identical genetic backgrounds; this is precisely why the concept of additive variance is so crucial in population genetics.

A different response, due to Dawkins (1982), argues that non-additive interaction is irrelevant thanks to particulate inheritance. Dawkins accepts that genes may 'blend' during embryology, that is, may interact epistatically in producing the phenotype, but notes that they do not 'blend' during transmission (p. 117). This response is also dubious, for it conflates the unit of inheritance with the unit of selection. Dawkins's point that genes are transmitted intact is of course correct. (Indeed he *defines* a gene as a length of DNA that is transmitted intact, i.e. that survives meiosis.) But to understand evolution, we need to ask *why* some genes are transmitted preferentially over others, which requires considering their phenotypic effects. If a gene's effect on phenotype is heavily dependent on the genetic background, understanding its dynamics from the gene's-eye perspective is made more difficult, particulate inheritance notwithstanding.

The issue of context-dependence is related to the famous 'beanbag genetics' controversy. The expression 'beanbag genetics' was coined by Mayr (1959) to denote the view that each gene is 'essentially... an independent unit, favoured or discriminated against by various causal factors', while organisms are 'atomistic aggregates of independent genes' (p. 614). Mayr attributed this view to Fisher, Haldane, and Wright, but as Wright (1960) showed, the attribution was largely unfounded.[12] Contemporary gene's-eye theorists are well aware that 'beanbag genetics' is mistaken, owing to the ubiquity of epistasis, so cannot fairly be accused of this position either.

Nonetheless, I think it is true that *if* beanbag genetics had proved correct, the language and metaphors used by gene's-eye theorists would be particularly natural. Thinking of genes as 'agents' with 'interests', capable of 'devising strategies' and 'manipulating the world around them', imputes to genes the ability to influence their own fate—an ability which epistatic interaction undermines. If it were possible to atomize an organism into a set of characters each controlled by one gene, with no epistatic interactions between genes, then the heuristic utility of the gene's-eye view would be even greater than it already is. So although gene's-eye theorists do not believe in beanbag genetics, they describe the world in a way that would be most suitable if it were true. Put differently, the gene's-eye perspective owes part of its heuristic utility to the fact that something like beanbag genetics often works to a first approximation.

Finally, it bears re-emphasis that while the heuristic utility of the gene's-eye perspective, in any given case, is partly a subjective matter, the level(s) at which selection takes place is not subjective. Epistasis and context-dependence may make the gene's-eye perspective less useful, but they have no bearing on whether selection is occurring, in whole or in part, at the genic level.

5.7 REDUCTIONISM AND PLURALISM REVISITED

In the light of the foregoing, we can now revisit the topics of pluralism and reductionism discussed in Chapter 4.

Recall that according to pluralists, the question 'at what level(s) is selection acting?' need not have a unique answer; different answers

---

[12] Though see Haldane's (1964) 'A Defence of Beanbag Genetics'. A good analysis of the beanbag genetics controversy is provided by de Winter (1997).

may be equally correct. Pluralism has often surfaced in the debate over genic selection and the gene's-eye perspective. For example, Buss (1987) regards the choice between a 'genic' and 'hierarchical' approach to the evolutionary transitions as a choice of language, not fact; Dawkins (1982) regards the gene's-eye view as 'ultimately equivalent' to the orthodox organismic view; while Sterelny and Kitcher (1988) defend 'pluralist genic selectionism', which says that any selection process can be represented in genic terms without loss of causal information.

If the foregoing arguments are correct, the true scope for pluralism in this area is easily determined. There is never room for a pluralist thesis about the *process* of genic selection. Whether a given evolutionary change is caused by genic selection, in whole or in part, is an objective matter. For example, if a gene spreads in a population by biasing segregation in its favour, then its spread is due to genic selection. Conversely, if a gene spreads by benefiting its host organism, but without biasing segregation or transposing to other loci within the genome, then its spread is not due to genic selection. Any 'representation' that suggests otherwise is simply mistaken about the causal facts.

With respect to the gene's-eye *perspective*, matters are different. If the above authors intend to assert that any selection process, at whatever level, can also be described from the gene's-eye perspective, then two responses are in order. First, this is not true, for reasons discussed in Section 5.5.1. Non-genetic inheritance, and focal units whose life cycles do not have bottlenecks, demarcate a class of evolutionary changes that cannot be captured in genic terms. Secondly, even where a gene's-eye view *can* be taken, this does not lead to pluralism in the sense of Chapter 4. For the gene's-eye and higher-level (or multi-level) descriptions answer different questions. As the bookkeeping argument (in its 'correct' version) shows, the former tells us which allele is fitter *overall*, and thus which will spread; the latter tells us the level(s) at which the causal process of selection is occurring. Once the process/perspective distinction is kept in mind, we can see that this is not pluralism about the levels of selection, as defined previously.

What about reductionism? In Chapter 4 we distinguished three senses of 'reductionism'. In the first sense, reductionism is the general injunction to explain properties of wholes in terms of their parts, that is, in a 'bottom up' way. In the second sense, specific to the levels-of-selection question, reductionism is the idea that lower levels of selection are explanatorily preferable to higher levels. In the third sense, also specific to the levels question, (indirect) selection at one level is 'reduced' to (direct) selection at a different level when there is a cross-level by-product running from the latter to the former.

The gene's-eye approach has often been described as 'reductionist', by both its detractors and supporters (cf. Williams 1985; Dennett 1995; Gould 2001). But in what sense? Not in the first sense. It is true that molecular biologists frequently give 'bottom up' explanations from a genic basis, for example, when they explain cellular differentiation by reference to which genes in the cell are being transcribed. But the molecular biologist's 'gene' is not the same as the 'gene' of the gene's-eye theorist, as Dawkins (1982) stresses. The former refers to a length of DNA that codes for a protein, the latter for a length of DNA that survives meiosis intact. So although the research programme of molecular biology is reductionistic, in the part–whole sense, this has no bearing on the gene's-eye approach to evolution (cf. Sarkar 1998).

What about the second sense? Does adopting the gene's-eye perspective commit one to saying that lower levels of selection are explanatorily preferable? The answer is 'no'. But a dialectical strategy employed by Williams (1966) has created the converse impression. Williams argued that selection at higher levels is intrinsically weaker than selection at lower levels, so should only be invoked if absolutely necessary; he also argued that 'genic selection' can explain all known adaptations. But this is a confusion. Williams was not using 'genic selection' in the sense recommended above; rather, he used it to mean selection at the organismic level that produces a change in gene frequencies. The thesis that selection is intrinsically more powerful at lower levels may or may not be true; but this has no bearing on the heuristic utility of adopting the gene's-eye view.

What about the third sense? Do gene's-eye theorists think that character-fitness covariance at higher levels is typically a by-product of, hence reducible to, direct selection at the genic level? Again the answer is 'no'; though again, the converse impression has been created. Williams's insistence on the distinction between adaptation and fortuitous benefit, combined with his view that 'genic selection' is all pervasive, might be taken to imply that all fitness-enhancing characters, of entities at all levels, are really there because they benefit genes. The same is true of Dawkins's (1976) claim that all adaptations are 'for the good of' the genes that cause them. But in the sense of 'genic selection' employed here, the reductionist thesis in question is false. Selection between the genes within an organism *might* have effects that filter up the hierarchy, generating spurious character-fitness covariance at the organismic level, or at some other level. But there is no reason to think that this is common, less still ubiquitous. For genic selection *itself* is relatively infrequent, thanks to the various organismic adaptations that suppress it.

So in what sense, if any, are gene's-eye theorists committed to reductionism? Obviously, they believe that genes exert a causal influence on phenotype, that is, that genetic differences are causally responsible, in part, for phenotypic differences. But this claim is wholly untendentious, at least for the vast majority of phenotypic traits. Where the traits in question are complex human behaviours, the claim has sometimes been criticized as 'reductionistic', but this is arguably an unhelpful usage. For adopting the gene's-eye view implies nothing about the relative importance of genes and environment in accounting for phenotypic differences.

Nonetheless, it is true that the gene's-eye perspective is at its most compelling when there is a relatively simple relation between genotype and phenotype, without too much epistatic interaction or context-dependence of genes' effects. The assumption that the genotype–phenotype relation is simple in this way might perhaps be called 'reductionistic'. Mayr (1959) used the expression 'holistic' to describe the converse assumption that epistasis is all-pervasive, and holism is often contrasted with reductionism. Similarly, Sarkar (1998) talks about 'reducing phenotypes to genes' in relation to the research programme of classical genetics. In this sense, gene's-eye thinking involves a weak commitment to reductionism, in that the heuristic utility of the gene's-eye view is greatest when the genotype–phenotype relation is at its simplest.

6

The Group Selection Controversy

INTRODUCTION

The group selection controversy is one of the most keenly contested chapters in the levels-of-selection debate. It also occupies centre stage historically, playing a role in the genesis of concepts such as inclusive fitness, selfish genes, evolutionarily stable strategies, and multi-level selection. The controversy shows no sign of abating today, as the divergent reactions to Sober and Wilson's *Unto Others* (1998) show.[1] In part, this reflects disagreement over empirical facts; but there are also conceptual and foundational issues at stake, which are the subject matter of this chapter.

In the light of the previous analysis of multi-level selection, it may seem odd that group selection merits treatment of its own. For multi-level selection rests on the idea that the Darwinian principles can be applied, in essentially the same way, up and down the biological hierarchy. So why should group selection—where a group is understood to mean a group of multicelled organisms—raise any special issues of its own?

The answer is that it should not. Indeed from the perspective of multi-level selection theory, the traditional contrast between 'group' and 'individual' selection is not wholly satisfactory. In traditional discussions, an 'individual' meant a multicelled organism, generally an animal. But as recent theorists have stressed, an organism is itself a group of cooperating cells, and a eukaryotic cell is a group containing organelles and the nuclear chromosomes. Moreover, to the extent that entities such as insect colonies are harmonious wholes, there is a case for regarding them as individual organisms.[2] This line of thought suggests

---

[1] See for example Maynard Smith (1998), Nunney (1998), Lewontin (1998), Skyrms (2002), and Dennett (2002).
[2] Queller (2000) argues that certain insect colonies should be regarded as organisms rather than superorganisms, precisely to make the point that 'organism' does not denote

that 'group' and 'individual' (or 'organism') are purely relative designations, much like 'collective' and 'particle' as used previously, rather than denoting fixed hierarchical levels as traditionally assumed.

Despite this consideration, I shall continue to use 'group selection' in its traditional sense, and to contrast it with 'individual selection'.[3] The historical importance of the group selection controversy is justification enough for this, even if it is not fully satisfactory from a theoretical standpoint.

Section 6.1 traces the origin of the group selection controversy, from Darwin to the present day. Section 6.2 looks at the distinction between MLS1 and MLS2 in relation to group selection. Section 6.3 examines the relation between group selection, kin selection, and evolutionary game theory. Section 6.4 explores a debate between Maynard Smith and Sober and Wilson over the status of trait-group models. Section 6.5 looks at what Sober and Wilson call the 'averaging fallacy'—which has led the importance of group selection to be obscured, they maintain. Section 6.6 examines the distinction between 'weak' and 'strong' altruism, and explores an argument of L. Nunney about how group selection should be defined. Section 6.7 revisits the issue of causal decomposition.

6.1 ORIGINS OF THE GROUP SELECTION CONTROVERSY

The origin of the group selection concept lies in Darwin's remarks on the evolution of worker sterility in social insects (Darwin 1859). Sterile workers do not reproduce themselves, but devote their whole lives to assisting the reproductive efforts of the queen. Such behaviour cannot easily evolve by individual selection, as Darwin realized, for it reduces individual fitness. But he suggested that colony-level selection might provide the answer. For colonies compete with other colonies; this could promote the evolution of traits that enhance colony fitness even if they are individually disadvantageous. Darwin reasoned similarly in his later speculations about the origin of altruism among hominids: 'a tribe including many members who . . . were always ready to give aid

---

an absolute level in the hierarchy; see Chapter 8, Section 8.1 for further discussion of this point.

[3] The label 'individual selection' is used instead of 'organismic selection' in this chapter, to accord with the terminology of traditional discussions.

to each other and sacrifice themselves for the common good, would be victorious over most other tribes; and this would be natural selection' (1871 p. 166).

The founders of modern neo-Darwinism—Fisher, Haldane, and Wright—each discussed group selection briefly. Fisher (1930) doubted its evolutionary importance, arguing that the turnover rate of groups is slower than that of individuals, so individual selection will predominate.[4] Haldane (1932) constructed a model to explore the evolution of 'socially valuable but individually disadvantageous characters', concluding that such characters could only evolve under special conditions (p. 49). Wright's views were somewhat different. His 'shifting balance' theory accorded a key role to differential migration between demes, as a means for the fittest demes to export favourable gene combinations to the rest of the species (Wright 1931). This constitutes group selection of a sort, as Wright realized; but it was not specifically addressed to the problem of altruism.[5]

Fisher and Haldane both emphasized that individual selection leads to the evolution of traits that benefit individuals, but not necessarily the groups or species to which they belong. Thus Haldane gave examples of adaptations that were 'biologically advantageous for the individual, but ultimately disastrous for the species' (1932 p. 65). Similarly, Fisher wrote that it was 'entirely open' whether the adaptations produced by natural selection 'in the aggregate . . . are a benefit or an injury to the species' (1930 p. 49).

In spite of this, certain mid-twentieth-century biologists routinely thought of evolution in terms of group benefit. Thus for example Konrad Lorenz (1963), discussing the adaptive significance of animal fights, explained the submissive behaviour of weaker animals as a feature designed to benefit the species. Similar arguments were made by Allee et al. (1949) in defence of the idea that animal groups are 'superorganisms'. Such arguments would have made sense if they had posited a process of group selection. But they did not: it was apparently assumed that group-beneficial outcomes would result from selection at the individual level. The notable exception was Wynne-Edwards (1962).

---

[4] Though Fisher allowed that species selection may have played a role in the evolution of sexual reproduction.

[5] Hodge (1992) treats Wright as an opponent of group selection, but other commentators regard the shifting balance theory as a precursor of modern group selection theory, e.g. Uyenoyama and Feldman (1980) and Wade (1978). See Wright (1980) for his last thoughts on group selection.

He too interpreted much animal behaviour as group-beneficial, but *did* posit a process of group selection to explain its evolution.

The 'good of the group' tradition came to an end in the 1960s, due largely to the work of Williams (1966). Williams argued that group selection was both inherently weak and theoretically superfluous—the phenomena it had been invoked to explain could be explained in more 'parsimonious' ways. Evolutionists should not appeal to group selection unless absolutely necessary, he insisted. The fragility of group selection as a mechanism for the evolution of altruism was also emphasized by Maynard Smith (1964); susceptibility to invasion by selfish cheats was its major weakness, he argued.

Williams stressed the difference between group adaptations,[6] which he thought are rare or non-existent, and 'fortuitous group benefits', which are common. Group adaptations are beneficial features of groups that evolved *because* they benefit groups, that is, by group selection. Fortuitous group benefits are beneficial features of groups that arose by other means. Selection on individual organisms is a possible source of fortuitous group benefits, Williams argued, for it can alter the composition of a group, thus enhancing group fitness. In Chapter 3's terminology, Williams is describing a cross-level by-product running from the individual to the group level.

As a result of Williams's and Maynard Smith's work, the concept of group selection fell into disrepute in the 1960s and 1970s. A consensus emerged that group selection was theoretically possible but empirically unlikely. This view was supported by mathematical models that suggested that group selection would only have significant effects for a limited range of parameter values (cf. Maynard Smith 1976). The influence of Hamilton's work on kin selection was also a contributory factor, for it apparently showed how altruistic behaviours could evolve without group selection, thus removing the motivation for invoking the latter. The same was true of Trivers's (1971) work on reciprocal altruism, and Maynard Smith and Price's (1973) work on evolutionary game theory.

In many evolutionary circles, the anti-group selection consensus is still intact. However, some contemporary theorists regard the wholesale rejection of group selection in the 1960s as a mistake. These neo-group selectionists do not advocate a return to the uncritical 'good of the group' tradition, but they think the pendulum has swung too far in the

---

[6] Williams uses the expression 'biotic adaptation' in place of 'group adaptation'.

opposite direction. This view has been most clearly put by D.S. Wilson and E. Sober (Wilson 1980, 1989, 2002; Wilson and Sober 1994; Sober and Wilson 1998).

What considerations underpin the neo-group selectionist revival? Wilson and Sober mention at least five. First, an exclusive focus on altruism has obscured the role of group selection in nature, they claim. Even if the conditions required for altruism to evolve by group selection are rarely met, it does not follow that group selection is an insignificant force. For group and individual selection might operate in the same direction, rather than opposing each other (cf. Wade 1978; Wright 1980).

Secondly, Wilson has long argued that his 'trait group' model provides a more plausible theory of group selection than the traditional models (Wilson 1975, 1977, 1980, 1983). (Recall the description of the trait-group model in Chapter 2 Section 2.3.2.) In traditional models the groups were taken to be demes—reproductively isolated, multigenerational, and geographically discrete. Wilson argues that many animals live in a different sort of group, often as a result of limited dispersal following birth. These 'trait groups' are the locus of ecological interactions. Trait groups are typically smaller than demes, and have a shorter duration. This helps circumvent one traditional objection to group selectionist explanations of altruism: that within-group selection will lead selfish types to fix. Periodic blending of the trait groups into the global population can prevent this (cf. Hamilton 1975).[7]

Thirdly and most controversially, Sober and Wilson (1998) argue that the supposed alternatives to group selection for explaining the evolution of altruism, such as kin selection and evolutionary game theory, are not in fact alternatives at all. Rather, they are *versions* of group selection, but presented in a formal framework which tends to obscure this fact. Clearly, this argument is conceptual rather than empirical; it turns on the meaning of 'group selection' and how it should be modelled.

Fourthly, Sober and Wilson suggest that group selection is especially plausible as applied to cultural evolution. One argument for this was discussed in Chapter 4—cultural variation will not necessarily be eroded by inter-group migration, as genetic variation typically is. A related argument is that the spread of selfish 'cheats' may be easier to prevent in human cultural groups than in animal groups, for humans'

---

[7] See Wade (1978) and Grafen (1984) for discussion of the contrast between traditional and intra-demic group selection.

cognitive capacities make detecting and punishing cheats easier. As a result, even large groups of altruists may be more resistant to invasion by cheats than genetical models have suggested.

Fifthly, Sober and Wilson allude to an argument from the literature on evolutionary transitions, which says that something akin to group selection must have occurred in the past. The evolution of multicelled organisms from single-celled ancestors, for example, required that free-living cells come together, curb their selfishness, and work for the good of the group. Since multicelled organisms clearly *have* evolved, the efficacy of group selection cannot be denied (Michod 1999). This particular example would not convince anyone who rejects the equation of kin and group selection, for kinship between cells helped drive the transition to multicellularity. But not all evolutionary transitions involved kinship. For example, eukaryotic cells evolved by the symbiotic union of unrelated prokaryotes, and chromosomes by the linking together of independent replicators (Margulis 1981; Maynard Smith and Szathmáry 1995). So it seems that cooperative groups can sometimes evolve without high relatedness between group members.

The revival of group selection described above is controversial. In part, this is because not all biologists agree that what is being revived *is* group selection in the traditional sense. A full assessment of the neo-group selectionists' case will not be attempted; but I try to comment on the main conceptual and theoretical points.

6.2 GROUP SELECTION AND THE MLS1/MLS2 DISTINCTION

Most discussions of group selection have a multi-level orientation: group and individual selection are considered simultaneously. This prompts the question: is the multi-level selection of the MLS1 or MLS2 variety? Recall that in MLS1, individuals are the 'focal' units and the two levels of selection contribute to a change in a single evolutionary parameter. In MLS2, individuals and groups are both 'focal' units, and the two levels of selection contribute to different evolutionary changes, measured in different currencies.

Many informal treatments of group selection, for example, Wynne-Edwards (1962), appear to describe an MLS2 process. Group selection occurs when some groups out-survive or out-reproduce others, where 'reproduction' means begetting new groups; as a result, some types of group become more common than others. However, most formal group selection models, traditional and recent, have an MLS1 structure; they aim to explain the evolution of an *individual* character, often altruism, in a group-structured population (Damuth and Heisler 1988). Thus Wade (1978), in a review of group selection models, observed that 'the genefrequency changes caused by group selection *(as is also true for individual selection)* consist of changes in the genetic composition of individuals within populations' (p. 102, my emphasis). So there is an element of mismatch between the informal discussions and the formal models.

This mismatch has often led to confusion. One such confusion concerns the role of extinction and colonization in traditional group selection models. In many of these models, selection occurs by some groups going extinct and others sending out colonists to found new groups (cf. Levins 1970; Boorman and Levitt 1973; Levin and Kilmer 1974). Colonization and extinction are analogous to organismic reproduction and death, fostering the illusion that the groups are playing a role isomorphic to that of individual organisms in models of individual selection. But as Heisler and Damuth (1987) observe, differential extinction and colonization, in traditional group selection models, are simply processes which contribute to group-level effects on individual fitness. The focal units are the individuals, not the groups.

This is the point missed by Maynard Smith (1976), who argues that differential extinction and colonization are pre-requisites of 'real' group selection. Maynard Smith arrives at this view by pursuing an analogy with individual selection. Just as individual selection requires individual death and reproduction, so group selection requires group death and reproduction, he reasons. So models of group selection that work by differential migration between groups, or by groups making differential contributions to a common mating pool, do not constitute real group selection, Maynard Smith thinks, for the groups themselves do not form parent–offspring lineages.

This would only be a good argument if the type of group selection in question were MLS2. But since virtually all formal group selection models, including those that operate via differential group extinction, have an MLS1 structure, Maynard Smith's argument is not a good one. All such models measure a group's reproductive success by how many offspring *individuals* it produces. So it is arbitrary to stipulate that group selection cannot be via differential migration/proliferation. (Indeed Uyenoyama and Feldman (1980) show that differential migration is a *more* effective mechanism for the spread of group-advantageous traits than differential extinction.) Maynard Smith might reply that 'real' group selection must be of the MLS2 variety, that is, must treat the groups as focal units, but this would imply that virtually none of the work purporting to be about group selection is actually about that subject.[8]

6.3 KIN SELECTION, RECIPROCAL ALTRUISM, AND EVOLUTIONARY GAME THEORY

The basic idea of kin selection was described in Chapter 5. A gene that causes its bearer to behave altruistically can spread by natural selection, so long as the cost to the bearer is offset by a sufficient amount of benefit to sufficiently closely related relatives. Hence Hamilton's famous rule $b > c/r$ for the spread of an altruism-causing gene.

How does kin selection relate to group selection? Many theorists regard them as fundamentally different evolutionary mechanisms, for example, Dawkins (1982), Trivers (1985), and Maynard Smith (1976). On their view, Hamilton's achievement was precisely to explain how altruism could evolve *without* invoking group selection. The striking empirical success of kin selection is presumptive evidence against the importance of group selection, they argue. But other theorists understand the relation between group and kin selection differently, as we shall see.

Though much altruism in nature is kin-directed, not all is: animals sometimes behave altruistically and cooperatively towards non-relatives, including non-conspecifics. Trivers's (1971) concept of reciprocal altruism provides one possible explanation.[9] Trivers noted that it may pay an individual to behave altruistically if there is an expectation of the favour being returned in the future. This requires that individuals interact on multiple occasions, and can recognize each other.[10] If individuals interact only once, there is no possibility of return benefit, so nothing to

---

[8] Moreover, it would disqualify Maynard Smith's own 'haystack model' from counting as group selection; but both he and others *have* regarded the haystack model as a type of group selection. See Maynard Smith (1964), Wilson (1987), and Sober and Wilson (1998) for discussion of the haystack model.

[9] Reciprocal altruism is a well-established theoretical idea, but its empirical significance is less clear. See Hammerstein (2003) for a recent discussion.

[10] This is true assuming a large population. However, if the population is so small that each individual interacts with only one or a few others, it is possible for reciprocal altruism to evolve without individual recognition, as Axelrod and Hamilton (1981) and Maynard Smith (1982) note.

Table 6.1. Payoffs for Player 1 and Player 2 in units of fitness

	Player 2	
	Cooperate	Defect
Player 1		
Cooperate	11, 11	0, 20
Defect	20, 0	5, 5

be gained by behaving altruistically. But with multiple encounters, and the ability to adjust one's behaviour according to who one is interacting with, altruism can evolve.

The well-known tit-for-tat strategy in the iterated Prisoner's Dilemma game illustrates these points. Players interact in pairs, and may adopt one of two possible strategies in any round: cooperate (C) or defect (D). Illustrative payoff values, in units of reproductive fitness, are shown in Table 6.1.

If the game is played only once, and pairing is random, then selection will favour defection, for the average fitness of defectors will exceed that of cooperators (Skyrms 1995). But if players meet repeatedly, and can adjust their strategy depending on their opponent's past behaviour, defection is not necessarily the fittest strategy. Axelrod and Hamilton (1981) show that if the probability of future encounters is sufficiently high, the tit-for-tat strategy yields a higher payoff than always defecting. In tit-for-tat, a player cooperates on the first round, and thereafter copies what his opponent did on the previous round.

Reciprocation apparently shows how altruistic behaviour can evolve without the need for group selection. This was Trivers's interpretation, and it is very natural. For reciprocal altruism, and the more general 'evolutionary game theory' of which it is part, is fundamentally a theory of individual advantage. The theory deals with situations where an individual's fitness is affected by its interactions with others, and analyses the strategy that maximizes individual fitness. No notion of group advantage is invoked. Indeed, Maynard Smith and Price (1973) devised evolutionary game theory to explain animal conflicts precisely because they were dissatisfied with Lorenz's 'good of the group' account.

Sober and Wilson argue that the relationship between group selection, kin selection, and evolutionary game theory has been widely misunderstood. They write: 'the theories that have been celebrated as alternatives to group selection are nothing of the sort. They are different ways of viewing evolution in multi-group populations... However, the theories are formulated in a way that obscures the role of group selection' (1998 p. 57). This is an intriguing claim; how do Sober and Wilson arrive at it?

Part of their argument derives from a point made by Hamilton in the 1970s that has not always been appreciated. The point is that what the evolution of altruism really requires is *a statistical tendency for the recipients of altruism to be altruists themselves*. Kin-directed altruism is the most obvious way of satisfying this requirement, but not the only way. Hamilton (1975) wrote: 'it obviously makes no difference if altruists settle with altruists because they are related... or because they recognize fellow altruists as such, or because of some pleiotropic effect of the gene on habitat preference' (p. 337). So kinship is just one way of generating the required statistical association between donor and recipient.

In one respect, Hamilton's claim that it 'makes no difference' whether altruists assort with each other because of kinship, or for other reasons, is misleading. Where altruism is directed at kin, the probability that donor and recipient share genes is the same *at every locus in the genome* (Grafen 1985; Maynard Smith 1976; Okasha 2002). So all the donor's genes benefit equally from the altruism—copies of them will be found in the recipient with greater than random chance. But where altruism is directed at non-relatives who just happen to be altruists, this is not so. Donor and recipient will both share the gene for altruism, but at all unlinked loci they are no more likely than average to share genes. So modifiers at other loci that suppress the altruistic behaviour will be selected (so long as they do not affect the chance of receiving altruistic benefits from others).[11] In the long term, this means that kin-directed altruism will be more robust. Nonetheless, Hamilton is obviously right that statistical association of altruists, not kinship per se, is what permits altruism to spread.

Once this point about statistical association is recognized, Sober and Wilson argue, the fundamental similarity between kin and group selection becomes clear. Recall Darwin's group selectionist scenario for the spread of altruism among hominids. Darwin envisaged a population subdivided into tribes containing varying proportions of altruists. Tribes with a higher proportion did better than those with a lower proportion, though within each tribe, selfish types enjoyed a fitness advantage.

---

[11] This qualification is crucial. Modifiers that suppress the altruism but also result in the organism not receiving altruistic benefits from others will not spread. It is for precisely this reason that the 'green beard' gene of Dawkins (1982) does not induce selection for modifiers that suppress its effects, as Ridley and Grafen (1981) note. See Okasha (2002) for further discussion of this point.

Had Darwin provided a formal treatment, he would have found that this mechanism only works if altruists tend to be clustered together; otherwise, within-tribe selection for selfishness predominates.[12] At root, therefore, Darwin's scenario is simply a way of generating a statistical association between altruists. Population subdivision, like kinship, is just a means to this end. This is why Sober and Wilson regard group and kin selection as ultimately equivalent, a view also endorsed by Hamilton (1975).[13]

Opponents of this view argue that group selection requires discrete groups, separated spatially or reproductively from other groups; while kin-directed altruism may occur in a single ungrouped population (Maynard Smith 1976, 1998). But Sober and Wilson reply that in the relevant sense of 'group', a group exists whenever there are fitness-affecting interactions between organisms. (Recall the 'interactionist' conception of part–whole structure.) This is the basis of Wilson's trait-group concept, in which the groups are temporary alliances of organisms that assort together for part of their life cycle (cf. Chapter 2, Section 2.3.2). Classical kin selection is just the special case where the trait groups are composed of interacting relatives.

Sober and Wilson (1998) argue that evolutionary game theory also involves a component of group selection, where the trait groups are interacting pairs. Consider again the Prisoner's Dilemma game above. There are three types of pair: {C, C}, {C, D}, and {D, D}. Within the mixed pair, D is fitter than C—the former gets 20, the latter 0. However, the {C, C} pair is fittest overall—it has a combined fitness of 22, versus 20 for {C, D}, and 10 for {D, D}. So the standard ingredients for multi-level selection are in place. Individual selection operates on fitness differences within groups (pairs), favouring D over C; while group selection operates on fitness differences between groups, favouring {C, C} over the others. The overall evolutionary outcome depends on the balance between the two levels of selection. It would be straightforward to describe this using the Price equation.

This multi-level description of the Prisoner's Dilemma may seem eccentric, given the 'individualist' orientation of evolutionary game theory, but it is quite instructive. Recall that if pairing is random

---

[12] Formally, this means that the distribution of group types must be 'clumped', i.e. have a higher variance than that of the binomial distribution that results from random group formation.

[13] Hamilton (1996) reports that he arrived at this view in part as a result of the Price equation.

and the game is played only once, defection spreads. In multi-level terms, this is because individual selection dominates group selection. For group selection to dominate, it is necessary that cooperators should tend to be clustered together. The tit-for-tat strategy in the iterated Prisoner's Dilemma is a way of ensuring that this happens—it causes the crucial statistical association between cooperators (altruists), which is absent in the one-shot game.[14] The multi-level description thus helps us understand the evolutionary dynamics.

In response to this argument, Maynard Smith (1998) accuses Sober and Wilson of using the term 'group' far too liberally. This may seem fair. However, Sober and Wilson's assimilation of reciprocal altruism and kin selection to group selection does not stem *solely* from a semantic decision about what to call a group. Rather, their point is that there is a fundamental commonality in the underlying evolutionary mechanisms. Kinship, group structure, and reciprocation are all ways of getting altruists to direct their benefits onto each other. Adopting an ultra-liberal definition of a group, then recognizing a component of selection at the group and individual levels, brings out this commonality.

Nonetheless, there is something paradoxical about Sober and Wilson's position. For evolutionary game theory was devised by theorists seeking to *avoid* appealing to group selection. The logic behind Sober and Wilson's argument is clear, but it leads to strange places.

How might this strangeness be alleviated? One option is to invoke pluralism, that is, to argue that kin selection and evolutionary game theory *can* be conceptualized in a multi-level way, but can equally be conceptualized as individual selection in a complex social environment. In Chapter 4, we saw that two ingredients are necessary for pluralism: (i) an MLS1 scenario, and (ii) indeterminacy of hierarchical structure. The former ensures that the evolutionary dynamics can be described in both single-level and multi-level terms; the latter makes plausible the idea that both descriptions are equally faithful to the causal facts.

Condition (i) is clearly satisfied, but the status of (ii) is less clear. Sober and Wilson see no indeterminacy—trait groups are real entities, picked

---

[14] Formally, this means that in the one-shot game, the distribution of pairs f({x, y}) is given by the binomial, i.e. f({C, C}) = p^2, f({C, D}) = $2pq$, and f({D, D}) = q^2, where 'p' and 'q' denote the overall frequencies of co-operation and defection in a given round. Tit-for-tat causes a clumped distribution, in which f({C, C}) > p^2 and f({D, D}) > q^2, i.e. organisms tend to play C when their opponent plays C, and similarly for D.

out by the criterion of fitness interaction. They thus resist pluralism, arguing that only the multi-level description is causally faithful—the single-level description illegitimately 'averages' across groups (cf. Section 6.5). This is the correct thing to say *if* one accepts the reality of trait groups. However, an alternative view is that trait groups occupy the grey area between 'real' collectives and mere aggregates of particles. In Chapter 4 we argued that there is no clear way to choose between the strict interactionist view and the view that hierarchical structure is sometimes indeterminate; both have points in their favour. If this is right, then pluralism is one possible resolution.

A different way to resist Sober and Wilson's conclusion is to argue that trait-group selection is fundamentally unlike 'traditional' group selection, so even if kin selection/evolutionary game theory can be assimilated to the former, this does not warrant co-classifying them with the latter. To be successful, this argument cannot just turn on the point that trait groups are unlike the demes of traditional group selection models; all parties accept this point. The argument must be that the causal structure of the selection process is relevantly different in the two cases. Precisely such an argument has been made by Maynard Smith (1987a), to which I turn next.

6.4 MAYNARD SMITH VERSUS SOBER AND WILSON ON GROUP HERITABILITY

Maynard Smith's argument is simple. The group selection question is about 'whether there are entities other then individuals with the properties of multiplication, heredity and variation, and that therefore evolve adaptations by natural selection' (1987a p. 121). In the process envisaged by Wynne-Edwards, the groups do possess these properties—for they are 'sufficiently isolated for like to beget like' (p. 123). But in Wilson's trait-group model, where the groups periodically blend into the global population, this is not so. A given trait group 'does not give rise to a group of the same composition', Maynard Smith says (p. 124). Therefore trait groups do not exhibit heredity, so cannot evolve adaptations, so cannot be 'units of evolution'.

Importantly, Maynard Smith is *not* denying that the trait-group model involves selection at the group level (though in an earlier 1976 paper he did deny this). Rather, his claim is that the mechanism at work in the trait-group model is (a) fundamentally unlike traditional group selection, and (b) incapable of producing group adaptations.

In reply to Maynard Smith, Sober (1987) wrote: 'I agree that the process Wynne-Edwards postulated involves groups that exhibit heredity. David Wilson's... groups require no such thing' (p. 133). But this difference is irrelevant, Sober argued, because group heredity (heritability) is not needed for evolution by group selection. He wrote: 'for natural selection to produce evolution, heritability of some sort is essential. But for *group* selection to cause evolution it is not essential that the heritability be *group* heritability' (p. 136, emphasis in original). So group selection can lead to the evolution of altruism—a group adaptation—*without* group heritability, Sober claims.

In their recent book, Sober and Wilson (1998) make a different reply to Maynard Smith. They agree with him that the evolution of group adaptations requires group heritability—the point that Sober (1987) denied. They also agree that for selection at any level to cause evolution, there must be heritable variation in fitness *at that level*. But trait groups *do* satisfy the heritability requirement, they claim—despite what both Sober and Maynard Smith thought in their 1987 exchange. For although trait groups periodically blend, we can still identify parent–offspring lineages among them, and thus consider the heritability of any group character. Blending means that trait groups have multiple parents—but this is simply the analogue of sexual reproduction. At the individual level, sexual reproduction reduces the heritability of individual characters vis-à-vis clonality, but does not make the concept inapplicable. The same is true at the group level, Sober and Wilson argue.

Therefore, there are three competing views of the role played by group heritability in models of group selection, summarized in Table 6.2.

Which of these views is correct? Despite their apparent incompatibility, I think each is partly correct. In Chapter 2, we saw that there are actually *two* notions of group heritability. Group heritability$_2$ measures the resemblance between parent and offspring *groups*; it is relevant to multi-level selection of the MLS2 type. Group heritability$_1$ measures the resemblance between a parent group and the set of offspring *individuals* that it produces; it is relevant to MLS1. Recall also that in MLS1, it is possible to write the total evolutionary change in terms of the *global* individual heritability, without mention of group heritability.

When Sober (1987) argued that in the trait-group model, what matters is that there should be *individual* heritability, not group heritability,

Table 6.2. The role played by group heritability

	Is group heritability necessary for groups to evolve adaptations?	Do trait-groups exhibit group heritability?	Do trait-group and 'traditional' group selection belong together?
Maynard Smith 1987a	yes	no	no
Sober 1987	no	no	yes
Sober & Wilson 1998	yes	yes	yes

he was partly right. For the trait-group model has an MLS1 structure, so group heritability$_2$ is irrelevant. As we saw in Chapter 2, in MLS1 it is not necessary for the groups to stand in parent–offspring relations at all, nor therefore for parent–offspring resemblance at the group level. So Sober's assertion is correct, if he means group heritability$_2$.

When Sober and Wilson (1998) argue that group heritability *is* in fact required in the trait-group model, they too are partially correct. In MLS1, group selection means some groups contributing more individuals than others to the next generation.[15] For this to produce an evolutionary response, there must be a resemblance between a group and the set of offspring individuals that derive from it. So Sober and Wilson's assertion is correct, if they mean group heritability$_1$.

However, Sober and Wilson (1998) do not distinguish the two types of group heritability; they appear to be talking about group heritability$_2$. They stress that parent–offspring relations between trait groups can be discerned, permitting us to ask whether parent groups tend to resemble their offspring *groups*. But this question is irrelevant, given that the groups are not the focal units in the trait-group model.

It follows that in one respect, Maynard Smith (1987a) and Sober and Wilson (1998) are both mistaken. Maynard Smith argues that trait groups can't be 'units of evolution', since they do not 'give rise to other groups of the same composition' (p. 123). Sober and Wilson reply that trait groups do give rise to groups with *similar* composition. But since the trait-group model is of the MLS1 type, what really matters is that a trait group should give rise to *a set of offspring individuals* of

---

[15] This assumes the Price approach to MLS1; the issue between the Price and contextual approaches does not matter for the moment.

similar composition. When this condition is satisfied, there may be a resemblance between parent and offspring groups too; but this is an incidental side effect.

This point is important because periodic blending affects the two types of group heritability differently. Blending greatly reduces the group heritability$_2$ from what it would be if groups reproduced asexually, for example, by fission. So in a model of the MLS2 type, blending limits the possible response to group selection. But blending has no effect on the group heritability$_1$. The resemblance between a parent group and the set of offspring individuals it produces will be exactly the same, whether or not those individuals blend with the progeny of other groups to form the next generation of groups.

To sum up: the Sober (1987) and Sober and Wilson (1998) positions are both correct, in different senses of group heritability. The dispute between Maynard Smith (1987a) and Sober and Wilson (1998) is misplaced; both parties miss the point that group heritability$_1$ is the relevant quantity in the trait-group model.

Where does this leave us? Maynard Smith argued that trait-group selection cannot produce group adaptations, and should not be co-classified with 'traditional' group selection. On the latter point, he is surely wrong. While it may be true that Wynne-Edwards was imagining an MLS2 process, most traditional group selection models were of the MLS1 type. So the fact that in the traditional models the groups were isolated demes that did not blend is beside the point; for the evolutionary mechanism was the same as in the trait-group model, just as Sober and Wilson say.

What about the first point? Since trait-group selection can lead to the evolution of altruism, which is group beneficial, surely Maynard Smith is wrong here too? But this might be disputed. For the trait-group model explains the changing frequency of different types of individual, *not* group. So while it is true that altruism is group-beneficial, and evolves for that reason, arguably it is an individual adaptation, for it is a property of individuals, not groups. The corresponding group property is 'proportion of altruists'; but group selection of the MLS1 type does *not* explain the evolution of that property, as stressed previously.

This issue was not discussed in previous chapters, for our focus was on selection rather than adaptation. How should it be resolved? On the one hand, the idea that group adaptations must be properties of groups, rather than individuals, seems natural. On the other hand, this severs the link between selection and adaptation, presuming we allow that MLS1 counts as genuine multi-level selection, for it implies that traits that evolve by selection at a given level need not be adaptations of entities at that level, which is counter-intuitive. I see no conclusive way of deciding the issue. Ultimately, how we use the expression 'group adaptation' does not matter greatly; what is important is to be clear about the logic of the underlying selection process.

6.5 THE AVERAGING FALLACY

I turn now to a different issue. Sober and Wilson (1998) argue that the importance of group selection in evolution has been obscured by what they call the 'averaging fallacy'. Those who commit this fallacy are prone to claim that certain selection processes involve only individual selection, when in fact they involve a component of group selection. Indeed, Sober and Wilson say 'the controversy over group selection and altruism in biology can be largely resolved simply by avoiding the averaging fallacy' (1998 p. 34).

The averaging fallacy occurs as follows. A population contains individuals of two types, living in groups. Overall (population-wide) fitness values for each type are calculated, by averaging across all the groups; it is then claimed that the type with the higher overall fitness spreads by individual selection. This is fallacious, Sober and Wilson argue, because it ignores the effects of group structure. The type that is fittest overall may actually be *less* fit within each group, if groups in which the type is common are fitter than groups in which it is rare. If so, then there is a component of both individual and group selection at work, they argue.

Sober and Wilson acknowledge that if we merely wish to predict the evolutionary outcome, averaging fitnesses across groups is fine. But if we wish to understand *why* one type has spread at the expense of the other, averaging is at best useless and at worst misleading. It is useless because it tells us nothing about the level(s) of selection; it is misleading because it invites the fallacious inference that the type that has spread has done so by individual selection, since its overall individual fitness must have been higher. To avoid the averaging fallacy, Sober and Wilson argue, it is essential to define 'individual selection' in terms of fitness differences *within* groups, à la Price, not averaged across groups.

The averaging fallacy is reminiscent of the 'bookkeeping objection' against the gene's-eye view, discussed in Chapter 5. According to the bookkeeping objection, gene's-eye theorists create the illusion that genic selection is the only causal force at work, by averaging the fitness of a gene across all the diploid genotypes in which it occurs. Similarly, opponents of group selection use averaging to create the illusion that individual selection is the only causal force at work, Sober and Wilson argue. So in both cases, averaging obscures the true level(s) of selection.

Precisely how close is this parallel? The answer depends on which version of the bookkeeping objection we are dealing with. In Chapter 5 we contrasted the original version, which focused on heterosis, with the 'correct' version, which focuses on meiotic drive. The key point, to recall, is that heterosis has no bearing on which level selection is occurring at, while meiotic drive does. The fact that the overall genic fitness values can be the same whether or not there is meiotic drive is thus the relevant consideration.

There is a perfect parallel between the averaging fallacy as described by Sober and Wilson and the 'correct' version, though not the original version, of the bookkeeping argument. Indeed, both arguments are instances of a point noted in Chapter 2. In an abstract multi-level scenario, consisting of particles nested within collectives, the change in average particle character can be expressed in terms of the global character-fitness covariance among the particles; alternatively, it can be partitioned into two components à la Price. The global covariance predicts the evolutionary change, but is silent on whether particle or collective-level selection, or both, are responsible. Those who commit the averaging fallacy in relation to group selection, and those gene's-eye theorists who attribute all evolutionary change to 'genic selection', are both taking the *global* character-fitness covariance to define particle-level selection. But this is a mistake, for it is tantamount to defining collective-level selection out of existence. This point is the essence of Sober and Wilson's (1998) discussion; it is also the essence of the 'correct' version of the bookkeeping objection.

Sober and Wilson are surely right that the averaging fallacy *is* a fallacy. The only circumstance in which this might be disputed is if there is indeterminacy of hierarchical structure; for then it becomes moot whether a multi-level or single-level description of the overall change is preferable. If one did not accept that a given population is hierarchically structured, that is, subdivided into genuine groups, one would see nothing wrong with averaging; indeed, one would regard the multi-level description as a statistical artefact. So for example, in the Prisoner's Dilemma game discussed above, where the 'groups' have a fleeting existence, someone might well argue that computing the overall fitness of each type is *not* just a computational convenience, but provides a better reflection of the underlying dynamics than the multi-level description. However, if the existence of hierarchical structure is not in dispute, then averaging fitnesses across groups and attributing all the change to 'individual selection' is indeed fallacious, for it obscures the role of group selection.

Importantly, the averaging fallacy can only be used to obscure group selection of the MLS1 type, not the MLS2 type. For in MLS2, group selection affects the evolution of a group character, not an individual character; so there is no way that an 'individualist' re-description is possible. This ties in with the point that pluralism is only a theoretical option in relation to MLS1, for only there is it possible to describe the overall change in both multi-level and single-level terms.

How common is the averaging fallacy? Sober and Wilson admit that in its 'general form' it has rarely been committed, but argue that particular instances are common (1998 p. 34). I agree with this assessment. Perhaps the most salient instance is the frequent suggestion that altruism in nature is 'only apparent'. Ghiselin (1974b) argued this way when he described altruism as a 'metaphysical delusion', as did Trivers (1971) and, arguably, Dawkins (1976). This idea arises naturally from averaging individual fitnesses across groups. For given that the individual-type, or genotype, with the highest overall fitness will spread, surely the behaviour of that type ultimately benefits it, hence is selfish? But this is a bad argument, as Sober and Wilson note, for it is tantamount to defining selfishness as 'whatever evolves'. Just as averaging threatens to define group selection out of existence, so it threatens to make 'the evolution of altruism' a contradiction in terms.[16]

To this point, our discussion of the averaging fallacy has largely agreed with Sober and Wilson. I turn now to an important challenge to Sober and Wilson due to L. Nunney (1985a, b, 2002). Like Sober and Wilson, Nunney agrees that 'individual selection' should not be defined in terms of differences in overall individual fitness, averaged across groups. However, he argues that Sober and Wilson's account of how individual selection *should* be defined, that is, in terms of fitness differences within groups, is flawed. Nunney's argument also brings out

---

[16] Uyenoyama and Feldman (1980) write: 'several authors have maintained that "evolution of altruism" is a contradiction in terms since it implies ultimately "selfish" promotion of the altruistic genotype by some means' (p. 381). The authors they cite include Haldane (1932), Trivers (1971), West-Eberhard (1975), and Dawkins (1976), but the list could easily be extended.

an ambiguity in the concept of biological altruism, and ties in with the discussion of contextual analysis from Chapter 3.

6.6 RANDOM VERSUS ASSORTATIVE GROUPING, STRONG VERSUS WEAK ALTRUISM

Before Nunney's argument can be addressed, some preliminaries are necessary. On the standard definition in biology, a behaviour counts as altruistic if it reduces an individual's fitness but increases the fitness of others. 'Fitness' here is usually interpreted as *absolute* fitness; so altruism entails a reduction in the donor's absolute fitness. This is the basis of Nunney's 'mutation test' for classifying behaviours. The mutation test compares the fitness of an individual that performs a given behaviour with what its fitness would be if it 'mutated' and stopped doing so. Only if its fitness would increase does the behaviour count as altruistic.

Sober and Wilson's concept of altruism is subtly different. They argue that a behaviour counts as altruistic if two conditions are met. First, within any group, individuals that perform the behaviour are less fit than those that don't; secondly, the greater the proportion of individuals that perform the behaviour, the greater the group's fitness.[17] These conditions do *not* imply that altruism reduces the donor's absolute fitness. For consider a behaviour that increases an individual's fitness by x, but increases the fitness of every other group member by y, where y > x > 0. The behaviour counts as altruistic by Sober and Wilson's lights, but fails the mutation test. If an individual who performs the behaviour ceased to do so, his fitness would be reduced, not enhanced. Behaviours of this sort, which boost an individual's fitness but boost that of others by even more, were termed 'weakly altruistic' by Wilson (1980). They contrast with 'strongly altruistic' behaviours that reduce absolute fitness, hence pass the mutation test.

The weak/strong altruism distinction can be expressed more precisely. Suppose individuals of two types, A and B, live in groups of size n; see Figure 6.1. Let $W_A(x)$ denote the absolute fitness of an A individual in a group containing x A types and (n-x) B types; similarly for $W_B(x)$. Let $G(x)$ denote the fitness of such a group, that is, its average individual fitness, so $G(x) = 1/n \, (x \, W_A(x) + (n - x) \, W_B(x))$.

To capture the idea that type A is altruistic, consider the following conditions:

[17] 'Group fitness' here means group fitness$_1$, i.e. average individual fitness.

	{AAAB}	{BBBB}	{AABB}	
{AAAA}	{AAAB}	{AAAB}	{ABAB}	{BBBB}
{ABBB}	{AABB}	{AABB}	{BBBB}	{ABBB}
	{AAAB}	{ABBB}	{AAAA}	

Figure 6.1. Individuals of two types in groups of size n = 4

(i) $G(x + 1) > G(x)$ for all x ('benefit' condition)
(ii) $W_B(x) > W_A(x)$ for all x ('cost' condition)

Condition (i) says that group fitness increases with the proportion of A types in the group. Condition (ii) says that within any group, A types are less fit than B types. However, (ii) does *not* imply that if a B individual mutated to A, he would suffer a fitness loss; it leaves open that a B→A mutant increases his own fitness, so long as he increases that of others by even more. Therefore, the conjunction of conditions (i) and (ii) implies only that the A type is *weakly* altruistic.

To define strong altruism, the same 'benefit' condition can be used, but a stronger 'cost' condition is needed:

(iii) $W_B(x) > W_A(x + 1)$ for all x ('cost' condition)

Condition (iii) says that the fitness of a B type in a group containing x A types is greater than the fitness of an A type in a group containing x + 1 A types; it captures the idea that a B→A mutant suffers an absolute fitness loss. (Note that if a B→A mutant starts out in a group containing x A types, after mutation he will be in a group containing x + 1 A types.) Therefore, the conjunction of (i) and (iii) says that the A type is *strongly* altruistic.[18]

The key difference between the two 'cost' conditions is that (ii) compares the fitness of A and B individuals within the same group, while (iii) compares the fitness of A and B individuals in different groups but with the same *neighbours* (Nunney 1985a). (An individual's neighbours are all the other individuals in its group.) To illustrate, consider an {ABBB} group. For (ii) to be satisfied, an A individual in this group must be less fit than a B individual in the same group. For (iii) to be satisfied, an A individual in an {ABBB} group must be less fit

---

[18] See Kerr, Godfrey-Smith, and Feldman (2004) for discussion of the various possible formal definitions of altruism and their interrelations.

than a B individual in a {BBBB} group, this being the group he would inhabit if he mutated.

Should weakly altruistic behaviours be included in the definition of altruism? Biologists have often disagreed on this question.[19] Wilson (1980) answered 'yes' on the grounds that weak altruism cannot evolve without group selection. In a single panmictic population, not subdivided into groups, weak altruism cannot evolve, for individuals that perform a weakly altruistic behaviour are less fit than those that do not. By defining altruism to include weak altruism, Wilson aimed to preserve the idea that behaviours favoured by individual selection are selfish, while those that require group selection to evolve are altruistic.

Although weak and strong altruism both require group structure to evolve, the former can evolve more easily. For if the groups are formed by random sampling from the global population, strong altruism cannot evolve, but weak altruism can. Random sampling means that the frequency distribution of groups containing x A types and (n - x) B types is binomial. So in the n = 4 case depicted in Figure 6.1, the frequency distribution of group types under random sampling is:

Group Type	{AAAA}	{AAAB}	{AABB}	{AABB}	{BBBB}
Frequency	p^4	$4p^3q$	$6p^2q^2$	$4pq^3$	q^4

where p and q are the overall frequencies of A and B in the global population.

Random group formation is not the only possibility. One alternative is 'positive assortment', that is, individuals tending to form groups with others of the same type. Positive assortment increases the between-group variance from what it would be under random group formation, thus increasing the power of between-group selection.

Wilson (1980) showed that with random group formation, weak altruism can spread—the between-group selection will be strong enough to dominate the within-group selection. But strong altruism can only spread if the groups are formed assortatively, for otherwise the between-group selection will be insufficiently strong. (See Section 6.7 for formal proof.) The reason is simple: the evolution of strong altruism requires that the benefits of altruism fall preferentially on other altruists, as

---

[19] See for example Wilson (1980), Nunney (1985a,b, 2002), Grafen (1984), and Uyenoyama and Feldman (1992). A useful survey of the issue is given by Kerr, Godfrey-Smith, and Feldman (2004).

discussed previously. This requires that group formation be assortative, rather than random.[20]

Interestingly, Nunney (1985a) argues that group selection itself, not just the evolution of strong altruism, requires assortative grouping. Prima facie, this argument is odd. Indeed, it seems as if Nunney must be conflating process with product. Group selection is meant to be a causal process in nature, the spread of strong altruism one of its products. In general, a causal process cannot be identified with one of its typical effects; typical effects do not have to occur. So although group selection has usually been invoked to explain the evolution of strong altruism, and although strong altruism can only evolve if groups are formed assortatively, it seems odd to argue that group selection *itself* requires assortative grouping.

The importance of not conflating process with product was a major theme of D.S. Wilson's (1980) book. There he argued that an exclusive focus on altruism has led evolutionists to underestimate the importance of group selection. For example, Williams (1966) discounted group selection on the grounds that adaptations that benefit groups but are costly for individuals are not to be found in nature. But as Wilson argues, group selection could occur without leading strong altruism to evolve. To think otherwise is to conflate process with product.

In fact, however, Nunney is *not* conflating process with product; his argument is more subtle. In part, Nunney's argument stems from rejecting the Price approach to MLS1. When Wilson (1980) argues that random group formation is compatible with group selection, he is relying on the Price approach — 'group selection' means between-group selection, and 'individual selection' means within-group selection. Given the Price approach, Wilson is correct. With random group formation, it is quite possible for a group character, for example, 'proportion of A types', to covary with group fitness; this will automatically be so if the A type is altruistic. But as we have seen, the Price approach is arguably flawed, for it ignores cross-level by-products. In effect, Nunney's claim is that with random group formation, the entirety of the group-level character-fitness covariance must be a by-product of individual-level selection.

Why should this be so? Even if we grant that the Price approach is flawed, that is, that the group-level covariance might be 'caused

---

[20] Here I am assuming that the groups break up and re-form every generation. If the groups stay together for more than one generation, then the crucial statistical association between donor and recipient could arise even with random group formation, owing to differential reproduction within groups (cf. Skyrms 2002, Fletcher and Zwick 2004).

from below', what has this to do with random group formation? To understand Nunney's reasoning, consider the scenario depicted in Figure 6.1. There are two factors that can affect an individual's fitness: its own character (A or B), and the characters of its neighbours (AAA, AAB, ABB, or BBB). As discussed previously, for there to be group selection, an individual's fitness cannot depend only on its own character, *pace* Price. But this condition is not sufficient, Nunney argues. Suppose that the A type is altruistic (weak or strong); so any individual does best if it has AAA neighbours, worst if it has BBB neighbours. Now suppose that group formation is random. This means that the benefit of having AAA neighbours *falls on A and B types equally*; similarly with the disbenefit of having BBB neighbours. Therefore, differences in individual character must be responsible for any differences in individual fitness, Nunney reasons, so individual selection is the only force at work. For owing to random group formation, the A and B types both experience each combination of neighbours with the same frequency; so if the A type spreads and the B type declines, for example, this can only be due to differences in individual character. Hence the conclusion that non-random grouping is necessary for group selection.

A natural corollary of Nunney's argument, which he draws, is that weakly altruistic behaviours are in fact selfish. Wilson argued the opposite in order to preserve the idea that traits evolvable by pure individual selection are selfish, while those that require a component of group selection to evolve are altruistic. Nunney also preserves this idea, but modulo a different understanding of individual and group selection. Weak altruism does *not* require group selection, Nunney argues, for it can evolve in a group-structured population with randomly formed groups, hence by individual-level selection. Strong altruism can only evolve with assortative grouping, so does require a component of group selection. Thus Nunney's argument goes hand in hand with the idea that the mutation test, rather than the comparison of within-group fitnesses advocated by Sober and Wilson, is the right way to define altruism.

If correct, Nunney's argument has significant implications for the neo-group selectionist revival. Sober and Wilson (1998) place considerable emphasis on the role played by female-biased sex ratios in restoring the credibility of group selection in the 1980s. Female-biased sex ratios are group adaptations, they argue—they enhance group fitness, even though within any group, individuals that produce a female-biased brood are less fit than those that do not (cf. Colwell 1981; Wilson and Colwell 1981). However Nunney (1985b) argues, on the basis of the mutation test, that female-biased sex ratios are often selfish traits—if an individual producing a female-biased brood ceased to so do, she would suffer an absolute fitness loss. Accordingly, female-biased sex ratios can evolve in a group-structured population with random group formation—hence by individual selection, according to Nunney.

Though the logic of Nunney's reasoning is clear, his conclusion is admittedly unusual. Sober and Wilson (2002b) try to rebut Nunney by exploiting the analogy between diploid population genetics and multi-level selection. As we have seen, a diploid organism can be thought of as a 'group' containing two alleles, permitting us to regard the overall gene frequency change in a diploid model as the result of two levels of selection: genic and organismic. Random group formation thus corresponds to random mating, which leads genotype frequencies to be in binomial, that is, Hardy-Weinberg, proportions; and assortative group formation corresponds to assortative mating. If Nunney were right that group selection requires assortative grouping, Sober and Wilson reason, parity of argument would imply that organismic selection requires assortative mating. But this is a most implausible conclusion. Population genetics models almost always assume random mating, for mathematical convenience, but no one thinks that this disqualifies the organism from being the unit of selection. Therefore group selection cannot require non-random group formation either.

This argument *looks* right, but is problematic for a reason discussed in Chapter 5, Section 5.4. We saw that in order to treat diploid population genetics as a multi-level system of the MLS1 type, the Price approach is superior to the contextual approach; but in relation to group selection it is the other way round. Nunney's thesis that with random group formation, all the selection is at the individual level is intimately related to rejecting the Price approach to disentangling the levels of selection; but in relation to diploid population genetics, the Price approach works well. Therefore, Sober and Wilson's rebuttal does not succeed. Accepting what Nunney says about group selection would only force us to endorse the parallel claim about diploid population genetics if the same approach to multi-level selection could be applied to both; and we know that it cannot.

This leaves open two questions. First, is Nunney's thesis correct? Secondly, how exactly does Nunney's approach relate to contextual analysis, which also stems from the idea that the Price approach to defining individual and group selection is flawed? I address these questions below.

6.7 CONTEXTUAL ANALYSIS VERSUS THE NEIGHBOUR APPROACH

Recall how contextual analysis works. It uses a multiple regression model to assess whether an individual's fitness is affected by individual character, group character, or both:

$$w = \beta_1 z + \beta_2 Z + e$$

where w is individual fitness, z is individual character, and Z is group character, defined for the moment as average individual character. β_1 and β_2 are the partial regressions of fitness on individual and group character respectively. Presuming the individual character is transmitted perfectly, this permits the overall change to be written as:

$$\begin{array}{rcl} & & \text{Group selection} \quad\quad \text{Individual selection} \\ \textit{Contextual partition}: \overline{w}\Delta\overline{z} & = & \beta_2 \text{Cov}(z, Z) \quad + \quad \beta_1 \text{Var}(z) \\ & = & \beta_2 \text{Var}(Z) \quad\quad + \quad \beta_1 \text{Var}(z) \end{array}$$

which constitutes a rival to the Price equation partition. The opposition between the contextual and Price approaches to MLS1 was examined in Chapter 3, Section 3.4.

To apply the contextual partition to the population depicted in Figure 6.1 above, assume that individuals reproduce asexually and that types breed true. We define $z = 1$ if an individual is A, 0 if it is B; so Z is the proportion of A types in a group. So for example, an A individual in an AABB group has $z = 1$ and $Z = 1/2$. Obviously, $\overline{z}$ is the overall frequency of the A type, and $\Delta\overline{z}$ the change in frequency over one generation.

Though Nunney (1985a) employs a different formalism, his approach is a close relative of contextual analysis, but with one difference. Whereas the contextual approach looks for 'group effects' on individual fitness, Nunney looks for 'neighbourhood effects' on individual fitness. In effect, Nunney is using a variant of the contextual regression model in which the second independent variable is neighbourhood character X, rather than group character Z. An individual's neighbourhood character is the average character of its neighbours, that is, of all the individuals in the group *except itself*.[21] So in the example above, an A individual in an {AABB} group has $X = 1/3$, while a B individual in the same group has $X = 2/3$. The regression model is:

$$w = \beta_3 z + \beta_4 X + e$$

where β_3 and β_4 are the partial regressions of fitness on individual and neighbourhood character respectively. Let us call this the *neighbour approach* to group selection. The overall evolutionary change can then be expressed as:

$$\text{Neighbour partition:} \quad \overline{w}\Delta\overline{z} = \underset{\text{Group selection}}{\beta_4 \text{Cov}(z, X)} + \underset{\text{Individual selection}}{\beta_3 \text{Var}(z)}$$

which constitutes a rival to both the Price and the contextual partitions. Of course, the neighbour, contextual, and Price partitions are all correct as *statistical* decompositions of the overall change, but at most one of them can be correct as a *causal* decomposition.

In motivation, the contextual and neighbour approaches are similar—both are intended as correctives to the Price approach—but they are subtly different. Consider the component of change that each attributes to group selection. On the neighbour approach, this component is β_4 Cov (z, X); on the contextual approach, it is β_2 Cov (z, Z). Crucially, Cov (z, Z) = Var (Z) so is always non-negative. But the term Cov (z, X) is very different. This term is the covariance between individual character and neighbourhood character, and may take on any value. If groups are formed at random, then Cov (z, X) will equal zero—there will be no correlation between an individual's character value and that of its neighbours. If groups are formed assortatively, then Cov (z, X) will be positive—an individual's character *will* correlate with that of its neighbours. Therefore, Cov (z, X) measures departures from randomness in the formation of groups.[22]

---

[21] Note that there is a simple relation between an individual's neighbourhood character X, its individual character z, and its group character Z, i.e. $X = 1/(n-1) (nZ - z)$. If follows from this that $\beta_2 = n/(n-1)\beta_4$, so there can only be 'group effects' on individual fitness if there are 'neighbourhood effects' on individual fitness and vice versa (cf. Okasha 2005).

[22] The contextual and neighbour approaches also embody different criteria for individual selection. On the contextual approach, individual selection requires fitness differences between individuals with the same group character, i.e. in the same group. But

This explains why Nunney regards non-random group formation as a prerequisite for group selection. If groups are formed randomly, then even if an individual's fitness *is* affected by which neighbours it has, that is, even if β_4 is non-zero, there will be no component of group selection, since β_4 Cov (z, X) will equal zero. By contrast, on the contextual approach, there *will* be a component of group selection, since β_2 Cov (z, Z) will be non-zero. Put differently, the contextual approach defines 'group selection' as direct selection on an individual's *group* character; while the neighbour approach defines 'group selection' as direct selection on an individual's *neighbourhood* character. In general, direct selection on a character y only affects the evolution of a character x if x and y are correlated. There is an *intrinsic* correlation between an individual's character and its group character (given how the latter is being defined); but there is only a correlation between an individual's character and its neighbourhood character if groups are formed assortatively. Hence Nunney's thesis that assortative grouping is necessary for group selection.

In Section 6.6 we noted that the evolution of strong altruism requires assortative grouping, while weak altruism does not. The neighbour partition yields a simple proof of this fact. Suppose type A is strongly altruistic. The first term of the neighbour partition, $\beta_3 \text{Var}(z)$, must then be negative. For β_3 measures the change in individual fitness if an individual with fixed neighbours has its z-value increased by one unit, that is, if a selfish individual converts to altruism and stays in the same group. By the definition of strong altruism, such an individual suffers a fitness loss, hence $\beta_3 < 0$. So if strong altruism is to evolve, that is, if $\Delta \bar{z} > 0$, then the second term of the neighbour partition, β_4 Cov (z, X), must be positive. Since random group formation implies Cov (z, X) = 0, it follows that strong altruism cannot evolve with random group formation. This conclusion does not go through for weak altruism, however, for weak altruism implies $\beta_3 > 0$. So for weak altruism, $\Delta \bar{z} > 0$ is compatible with Cov (z, X) = 0.

In Chapter 3, we argued that the contextual approach is theoretically superior to the Price approach, as the latter detects group selection even when individual fitness depends only on individual character. However,

on the neighbour approach, individual selection requires fitness differences between individuals with the same neighbourhood character; such individuals *cannot* be in the same group. A consequence of this is that on the neighbour approach, individual selection can operate even if there is no within-group variance in fitness, something that is impossible on both the contextual and Price approaches. See Okasha (2004b) for discussion.

this consideration does not support the contextual approach over the neighbour approach; for they both yield the 'correct' answer in the case that the Price approach gets wrong.

How then should we choose between the contextual and neighbour approaches? One possible argument for the latter stems from a general consideration about the likely causal influences on individual fitness. Since an individual interacts directly with its neighbours, not with its neighbours-plus-itself, neighbourhood character is the *sort* of character that could directly influence an individual's fitness, while group character is not. If an individual's fitness is affected by interactions with its neighbours, we should expect a correlation between individual fitness and neighbourhood character, and thus between individual fitness and group character;[23] but arguably only the former correlation is causal. So the neighbour approach yields a more satisfactory causal decomposition.[24]

This argument in favour of the neighbour approach seems right, *given that group character is defined as average individual character*. There seems no way that *this* group character could affect individual fitness directly, rather than via neighbourhood character. But as we saw in Chapter 3, the contextual approach is also applicable to 'emergent' group characters, which are not the average of any individual character. Where emergent characters are at issue, it is less clear that their causal influence on individual fitness must be indirect, mediated by neighbourhood character. For the set of neighbours of a given individual is not a 'real' biological unit the way a group is, so any given emergent character may not even be well-defined on this set. Therefore, where individual fitness is affected by emergent group characters, the contextual approach provides the better causal decomposition; the neighbour approach may not even be applicable.

This suggests that the neighbour approach is most plausible during the early stages of the evolution of cooperation/altruism, where cohesive,

[23] Here it is important to remember that β_2 and β_4 always have the same sign; see note 21.
[24] Importantly, there can be no purely statistical solution to the problem of whether group character Z or neighbourhood character X causally affects individual fitness. One might think that this question could be resolved by including *both* Z and X, along with z, as independent variables in a regression analysis, with fitness as response variable. But this will not work. Since any one of the variables {z, Z, X} is a linear function of the other two, the relevant partial regression coefficients will not be well-defined. (This is known as 'the problem of perfect collinearity'.) So extra-statistical considerations concerning the actual mechanisms by which neighbourhood and group characters might causally affect fitness are the only way of choosing between the approaches.

integrated groups have not yet evolved, but where individuals do engage in social interactions. During such stages, an individual's fitness may well be affected by factors such as the proportion of altruists in its neighbourhood. This ties in with a point stressed by Nunney (1985a), that the neighbour approach is perfectly applicable even if discrete, non-overlapping groups do not exist in the population. But where cohesive groups *have* evolved, with discrete boundaries and emergent properties of their own, the contextual approach makes more sense, for it is likely that individual fitness will be affected by emergent group characters. Still later, groups may be so cohesive that we wish to treat *them* as the focal units, that is, to move to an MLS2 framework. The idea of a transition between MLS1 and MLS2 is discussed in Chapter 8.

What then of Nunney's thesis about non-random group formation, and the correlative thesis that weak altruism is really a form of selfishness? I think Nunney is partially right. There are situations, such as those described by the simplest models for the evolution of altruism of the sort used above, where the neighbour partition seems 'better' than the contextual or Price partitions, in the sense of yielding a more accurate causal decomposition of the overall change. And if the neighbour partition is taken to define individual and group selection, Nunney's thesis follows immediately. However, there are other situations, typically involving cohesive groups rather than just interacting individuals, where the contextual approach seems superior; and modulo that approach, group selection does not require non-random group formation. If this is correct, then a general verdict on Nunney's thesis is not possible; it needs to be assessed on a case-by-case basis.

7

Species Selection, Clade Selection, and Macroevolution

INTRODUCTION

This chapter examines selection at the level of the species, a much-discussed idea in the macroevolutionary literature. Section 7.1 traces the origin of the species selection concept. Section 7.2 explores the debate over how 'real' species selection should be distinguished from its surrogates. Section 7.3 examines an argument of J. Damuth (1985), which says that species are not the right *sorts* of entity to be units of selection. Section 7.4 looks at clade selection, regarded by many as a natural generalization of species selection, and argues that it is conceptually incoherent.

7.1 ORIGINS OF SPECIES SELECTION

Species selection came to prominence in the 1970s through the work of Stanley (1975) and Eldredge and Gould (1972).[1] These authors sought to 'decouple' macro- from microevolution, arguing that the long-term evolutionary trends revealed in the fossil record are not simply side effects of microevolutionary processes, as was traditionally assumed, but the product of autonomous macroevolutionary forces, such as species selection. The basic idea of species selection is simple. If species vary, and the variation gives rise to differential extinction/speciation, then some

---

[1] Though Gould (2002) argues that de Vries (1905) was the originator of the species selection concept. Other precursors of the modern discussion include Fisher (1930), who thought that species selection might have played a role in the evolution/maintenance of sexual reproduction, and Dobzhansky (1951), who argued that intra-specific genetic variation was an adaptation of species, permitting rapid evolution in response to environmental stress.

types of species will become more common than others. Therefore, species themselves are the 'focal' units in species selection theory (cf. Arnold and Fristrup 1982).

Eldredge and Gould (1972) presented species selection as a corollary of punctuated equilibrium theory, which says that most morphological evolution happens during the process of speciation; thereafter, species remain in a state of 'stasis'. Evolutionary history thus exhibits a punctuated pattern: millions of years of stasis broken by episodes of rapid evolution, concurrent with the formation of new species by cladogenesis, or lineage splitting.

If punctuated equilibrium is the norm, what explains long-term evolutionary trends, such as the increase in body size in many mammalian lineages, for example, horses, over geological time? Traditional neo-Darwinism attributed these trends to directional selection within species. But if most species are in stasis, this explanation cannot be right. An alternative is that the trends were due to the morphological changes occurring at speciation. When an ancestral horse species split, for example, it might have produced daughter species that were larger than it (a type of 'directed mutation' at the species level). This is conceivable, but Eldredge and Gould argued that it is false. The morphological changes that occur at speciation are 'random' with respect to long-term trends, they claimed; a given horse species was as likely to have produced a smaller as a larger daughter. This generalization was dubbed 'Wright's rule', after S. Wright's views on speciation.

Widespread stasis combined with Wright's rule leaves just one explanation of evolutionary trends, Eldredge and Gould argued: species selection. If large horse species survive better than small ones, or leave more daughter species, and if body size is heritable at the species level, then over time there will be a trend towards increased size in the horse clade, despite the absence of any within-species change, and despite the 'randomness' of the morphological changes arising at speciation. Gould and Eldredge (1977) expressed this in the slogan 'punctuated equilibrium + Wright's rule = species selection'.

In their early work, Eldredge and Gould argued that species selection *depends* on both the ubiquity of stasis and the truth of Wright's rule.[2] But there seems no reason why this should be. Even if most species did

---

[2] Gould and Eldredge (1977) say that Wright's rule is a 'precondition' for species selection (p. 148).

undergo phyletic transformation, contrary to punctuated equilibrium, species-level selection could still operate, amplifying or counteracting the within-lineage changes (cf. Williams 1966). Similarly, even if Wright's rule were false, species selection could still occur—directed mutation does not imply the absence of Darwinian selection. In later work, Gould (2002) accepts that species selection is logically independent of these other aspects of punctuated equilibrium theory.

Both conceptually and historically, species selection is related to the 'individuality thesis' of Ghiselin (1974a) and Hull (1978), which says that species are individual entities, extended in space and time, rather than classes of organisms sharing a common property. Gene flow is the 'glue' that binds together the parts of a species into a genuine whole, on this view. The individuality thesis makes it plausible to think of species as entities with life cycles, that is, that are born, reproduce, and die, and thus as the right *sorts* of entity to function as units of selection. It is no accident that most advocates of species selection also endorse the Ghiselin/Hull view of species.[3]

Despite the plausibility of the individuality thesis, species are obviously not functionally organized the way other biological individuals, for example, cells, organisms, and insect colonies, are. These entities exhibit a division of labour between their parts, and have mechanisms for suppressing conflict among the parts, ensuring they work for the good of the whole. The same is not true of species. However, this does not necessarily invalidate the species selection concept. Since the turnover rate of species is much slower than that of colonies, organisms, and cells, it is unlikely that there has been sufficient time for species selection to produce comparably complex adaptations, even if it has operated.

Most biologists accept that species selection is possible, but there is considerable disagreement over its empirical significance.[4] However, the issue is not solely empirical. There are also disagreements over exactly what species selection amounts to, what types of phenomena it might explain, and what it means for macroevolution to be 'irreducible' to microevolution. It is these conceptual issues that will occupy centre-stage here.

---

[3] Though Williams (1992) is an exception.
[4] See for example Maynard Smith (1983), Dawkins (1982), and Gould (2002) for opposing views on this matter.

7.2 GENUINE SPECIES SELECTION VERSUS 'CAUSATION FROM BELOW'

It is clear that species do differ in their fitness, that is, their rate of survival/speciation, but this does not necessarily imply a process of species-level selection. The extraordinary speciosity of Hawaiian drosophilids is apparently due to the physical environment of Hawaii, which is especially conducive to speciation, rather than to any biological properties of the species themselves; so this is not species selection (Hoffman and Hecht 1986). Similarly, Raup et al. (1973) argue that although extinction rates do vary among species, the variation is largely random; so again, variance in species fitness is not indicative of species selection.

Genuine species selection requires that differences in species fitness be caused by differences in a species character, rather than arising for some other reason. This causal ingredient is present in certain putative examples. Recall Jablonski's hypothesis that the average geographic range of late-Cretaceous mollusc species increased as a result of species selection, discussed in Chapter 2. Jablonski (1987) argues that species with larger geographic ranges had a greater tendency to speciate, that is, that differences in range caused differences in fitness. Further, he argues that geographic range was a heritable character, so selection on it produced a cross-generational response.

Conceptually this may seem unproblematic, but in fact there are substantive disagreements over how causality at the species level should be understood. Below I examine a number of proposals for how to distinguish genuine species selection from other causal processes that can lead to variance in species fitness.

In a series of publications, E. Vrba has argued that true species selection is much rarer than its advocates have thought, if indeed it has occurred at all (Vrba 1984, 1989; Lieberman and Vrba 1995). Most of the alleged examples are suspect, for they do not involve 'causation at the focal level of species' (1989 p. 130). Vrba illustrates this point with the example of two African antelope clades, discussed briefly in Chapter 3. Recall that ecological specialists (stenotopes) tended to speciate more frequently than generalists (eurytopes), resulting in an evolutionary trend. But the trend was a side effect of organism-level processes, according to Vrba, which produced greater local differentiation in the stenotopic species, thus restricting gene-flow and enhancing the probability of speciation. So Vrba categorizes this as 'effect macroevolution' rather than genuine species selection.[5]

According to Vrba, an 'acid test' for genuine species selection is that it must in principle be able to oppose selection at lower hierarchical levels, though it need not do (1989 p. 115). This is an attempt to capture the idea that a higher-level selection process requires 'autonomy' from selection at lower levels. Interestingly, Vrba regards her acid test as intimately linked to the emergent character requirement. If the species character than covaries with species fitness is aggregate rather than emergent, then the required autonomy from lower-level selection cannot obtain, she claims, so species selection cannot be the right verdict.

How should these ideas be assessed? The acid test requirement seems correct. In general, the direction of selection can be different at different hierarchical levels. So it has got to be *possible* for selection at the species level to counteract the effects of organism-level selection, as Vrba says. But this has nothing to do with emergent characters. As we saw in Chapter 3, whether a given character-fitness covariance is a by-product of lower-level selection is independent, in both directions, of whether the character in question is emergent. So Vrba's 'acid test' is correct, but she is wrong to link it to the emergent character requirement (cf. Grantham 1995; Stidd and Wade 1995).

To illustrate this point, consider the hypothesis that species selection was involved in the evolution/maintenance of sexual reproduction. The rationale for this hypothesis is that selection at the organismic level should favour asexuality, because of the well-known 'two-fold cost of sex' (Maynard Smith 1978). But sexuality might be advantageous at the species level, permitting a more rapid evolutionary response to environmental stress, so species selection could have favoured the sexual over the asexual lineages. This hypothesis clearly satisfies Vrba's acid test—higher- and lower-level selection oppose each other. But the species character subject to selection—'contains sexually reproducing organisms'—is presumably aggregate rather than emergent.

In Chapter 3, we argued that the emergent character requirement stems from conflating the question whether lower-level *selection* is responsible for higher-level character-fitness covariance, with the question whether *some lower-level processes or other* are responsible. The

---

[5] Effect macroevolution takes its name from Williams's (1966) distinction between adaptation and fortuitous side effects.

former, not the latter, is the salient question; for on plausible metaphysical assumptions the answer to the latter will always be 'yes'. The distinction between effect macroevolution and species selection, as Vrba draws it, commits this conflation. For Vrba diagnoses effect macroevolution wherever organism-level processes, of whatever sort, can be shown to be 'ultimately' responsible for the differential speciation/extinction. But taken to its logical conclusion, this threatens to make species selection impossible; it is no surprise that Vrba comes close to saying just this.

It follows that at least some of Vrba's examples of effect macroevolution should actually be classified as species selection, including the African antelope example. In that example, Vrba admits that the differences in species fitness are not side effects of differences in organismic fitness; on the contrary, 'organismal and species success are, to a large extent, independent' (Vrba and Gould 1986 p. 224). So the reason that stenotopic species were fitter than eurytopic ones is *not* that the former contained fitter organisms than the latter. Thus the evolutionary trend towards stenotopy was not a by-product of organismic *selection*; though it is presumably true that some causal processes at the organismic level were 'ultimately' responsible. So this is a bona fide case of species selection.

The same applies to the example of early-Tertiary marine gastropods, regarded as genuine species selection by some (Hansen 1983; Stanley 1979; Gilinsky 1986; Jablonski 1982, 1986); but not by others (Lieberman 1995; Eldredge 1989). Larval development in gastropods can either be planktotrophic (i.e. the larvae feed in the plankton), or not. The fossil record shows an increased frequency of species with non-planktotrophic larvae over time. This was apparently due to limited dispersal in non-planktotrophs, which restricts gene flow and hence raises the probability of speciation. Again, this *should* be classified as species selection. Differences in species' fitness were *not* caused by differences in the fitnesses of their constituent organisms, but by differences in the extent of within-species gene flow. These differences in turn stemmed from differences in mode of larval development, so can ultimately be explained by organism-level processes. But crucially, the lower-level explanation is not a *selective* explanation.

I turn now to a different proposal for how to distinguish 'real' species selection from its surrogates. A number of authors have suggested that the distinguishing mark lies in differential speciation versus differential extinction. For example, Grantham (1995) argues that 'the concept of "speciation rate" cannot be expressed at the organismic level'; so cases of differential speciation are not 'reducible' to the organismic level, while cases of differential extinction are (p. 309n). Similarly, Gould (1982) argues that true species selection is more likely to be by differential speciation, because 'propensity to speciate is not generally a property of individuals' (p. 92); see also Gilinsky (1986) and Sterelny (1996a).

This suggestion seems incorrect. In general, selection at any level can operate either on differences in survival or fecundity; it would be odd if the species level were an exception. Moreover, there are plausible examples of species selection by differential extinction, for example, the hypothesis that species selection favoured sexual reproduction, mentioned above. This hypothesis says that sexual lineages had better survival prospects than asexual ones, *not* that their intrinsic rate of cladogenesis was higher. So while it is true that extinction occurs when all the organisms in a species die, and so in that sense can be 'expressed in organismic terms', this does *not* mean that a species' probability of extinction is solely a function of organismic fitnesses, which is the critical issue.

Empirically, it *may* be true that differential species extinction is often a side effect of organismic selection. For example, in cases of competitive exclusion, where two species compete for resources and one drives the other extinct, the differential extinction is likely to be a side effect of the fitness disadvantage suffered by organisms in one species vis-à-vis organisms in the other. Conversely, most cases of differential speciation are probably not side effects of selection on organisms. There seems no particular reason why a species whose constituent organisms are especially fit should be more or less likely to speciate as a result. Evolutionists have discussed various mechanisms of speciation, none of which imply a link between a species' probability of speciating and the fitnesses of its constituent organisms. However, such a link cannot be ruled out a priori.

To conclude, it may be true that differential speciation is a more promising outlet for genuine species selection, but not for the reasons often alleged. The point is not that speciation is 'something that happens to species' rather than to organisms, nor that extinction rate is 'expressible in organismic terms' while speciation rate is not. Rather the point is that, empirically, a species' probability of speciating is unlikely to be affected by the fitnesses of its constituent organisms, while its probability of going extinct may very well be. So 'causation from below', as that notion was explicated in Chapter 3, is more likely to be the correct verdict in cases of differential extinction, for empirical reasons.

7.3 SPECIES VERSUS AVATARS: DAMUTH'S CHALLENGE

Damuth (1985) argues that the idea of species selection is conceptually flawed, for species are not the right sorts of thing for selection to act on. He notes that most species are subdivided into a number of local populations, which can be widely distributed geographically. As a result, species do not interact with each other, or with the environment, in an ecologically meaningful way, so cannot be subject to selection. Of course, Damuth does not deny that rates of speciation and extinction vary; his claim is that this variation does not reflect a causal process of selection at the species level.

In place of species selection, Damuth recommends the concept of *avatar selection*. An 'avatar' is a local population of conspecific organisms, integrated into a particular ecological community. So an avatar of a baboon species, for example, might consist of all those baboons inhabiting a particular rain forest. The avatars within a community do interact and compete with each other, so can be subject to a selection process, Damuth argues.

One argument against Damuth's proposal is that avatars are much less well-defined than species. If the world's biota were divided into discrete ecosystems, between which there was little flow of energy or matter, avatars could be quite easily picked out, or individuated. But the world is not like this, which is why the 'reality' of ecosystems is a controversial issue in ecology. By contrast, species can be picked out by the criterion of reproductive isolation in a reasonably determinate way.

This means that fitness is more clearly defined for species than for avatars. Damuth sees no difficulty in treating avatars as fitness-bearing entities, since 'avatars may go extinct or produce other avatars' (1985 p. 1136). But since the local populations of a species normally exchange migrants to some extent, and sometimes fuse with each other completely, it is not obvious how avatar identity is to be judged. What determines when one avatar has become two? For species, by contrast, the onset of reproductive isolation provides a fairly clear criterion for when one species has become two, though there are problem cases.

The choice between species and avatars as focal units reflects the tension between the 'replicator–interactor' conception of evolution and Lewontin's 'heritable variation in fitness' conception, discussed in Chapter 1. For in effect, Damuth's point is that species are not 'interactors' in the sense of Hull (1981), while avatars are. This may be true. But on the other hand, fitness and reproduction are better defined for species than for avatars, so the Lewontin criteria apply more naturally to species. Previously we argued that the Lewontin approach is theoretically superior to the Dawkins/Hull approach. If this is right, then Damuth's revisionist proposal should not be endorsed. It is true that avatars come closer than species to being 'interactors' in Hull's sense; but the fundamental requirement for an entity to be a unit of selection is that it be capable of reproduction, hence form determinate parent–offspring lineages.

In reply to Damuth, Sterelny (1996a) makes the point that selective forces need not be spatially local. At the organismic level this is clear. For example, conspecific organisms are often affected by the same parasites wherever they are found, so inhabit a similar selective environment in that respect. So if there is selection for parasite-resistance, the organisms subject to selection may not be in direct competition, nor in physical contact with each other. The same is true of species selection. Even though species are often distributed across many separate communities, selection can still act at the species level. So long as species vary with respect to a character that causally affects species fitness, then species selection can operate; the selective forces need not be spatially local. Indeed, Sterelny argues that the property of being geographically widespread, or fragmented into many local populations, might *itself* be a character that affects a species' probability of extinction/speciation.

Damuth's argument is reminiscent of an earlier argument due to Fisher (1930). Fisher argued that species selection is unlikely to be a significant force due to 'the small number of closely related species which in fact do come into competition' (p. 121). But Sterelny's point about the non-locality of selective forces also shows that Fisher's argument is mistaken. Neither competition nor direct contact is needed for a set of species to be subject to a common selection pressure, any more than it is necessary for organisms. All that is necessary, at both levels, is that character differences should cause fitness differences.

The example of sexual reproduction illustrates this point.[6] Suppose it is true that asexual lineages have gone extinct because of their reduced capacity to evolve. The set of species whose composition was modified

---

[6] Ironically, Fisher himself allowed that species selection for sexual reproduction might have occurred; see footnote 1 on page 203.

by species selection would then include all those species, sexual and asexual, which faced environmental stresses such that the capacity to evolve affected their survival prospects. Clearly, most of these species will never have come into contact, and may be found anywhere on earth. For environmental deterioration is not restricted to a particular area; nor does it only affect certain taxa. So the species in question would have been geographically widespread and taxonomically diverse.

7.4 THE CONCEPT OF CLADE SELECTION

I turn now to *clade selection*, discussed by authors including Williams (1992), Sterelny (1996a), Vermeij (1996), and Nunney (1999). A clade is a monophyletic group of species, that is, a group comprising an ancestral species, all of its descendent species, and nothing else. Clades are thus located further up the (genealogical) hierarchy than species.[7] Clade selection is often presented as a natural generalization of species selection. Thus, for example, Williams (1992) writes: 'there is no reason why species selection should be recognized as a special process different from any other kind of clade selection . . . selection can take place among clades of higher than the species level' (p. 125). Similarly, Nunney (1999) says that species selection 'can be subsumed under the more general category of clade selection' (p. 247).

This idea may seem plausible, particularly if one endorses the view that all monophyletic groups, not just species, are 'individuals' in the Ghiselin/Hull sense. But in fact it faces a critical problem, for the notion of clade fitness turns out to be incoherent.

The fitness of a species is normally defined as the number of offspring species it leaves.[8] This notion makes sense because species reproduce,

---

[7] According to standard cladistic usage, followed here, a single species does not count as a clade and cannot be or fail to be monophyletic. Monophyly is a property of collections of species, not single species (Hennig 1966). However, some theorists have tried to apply the concept of monophyly to single species, usually by arguing that a species is a monophyletic group of populations (cf. Mishler and Brandon 1987). This goes hand-in-hand with a broader use of 'clade', according to which single species do count as clades.

[8] Some authors define species fitness slightly differently, as the difference between speciation rate and extinction rate, by analogy with the Malthusian parameter (e.g. Vrba 1984). A similar definition of organismic fitness is used by Michod (1999). The argument below—that there is no coherent notion of clade fitness comparable to the notion of species fitness—applies whichever definition of species fitness we prefer.

that is, beget other species. But do clades reproduce too? Proponents of clade selection believe that they do. Williams (1992) explicitly describes cladogenesis as reproduction for clades (p. 52). Similarly, van Valen (1988) argues that supra-specific taxa can beget other supra-specific taxa, hence be subject to selection; he talks about the probability that 'one family gives rise to another' (p. 59).[9] Sterelny (1996a) apparently concurs. He argues that clades have adaptations, and he insists that adaptations must be *heritable* characters, so they can be honed by cumulative selection. He rules out some alleged clade adaptations on the grounds that the characters in question are unlikely to be heritable; Vermeij (1996) makes a similar argument. Heritable means transmitted from parents to offspring, so Sterelny and Vermeij presumably think that clades can beget other clades.

However, there is a complication. For clades are by definition monophyletic, and as a matter of logic, monophyletic clades cannot stand in ancestor–descendent relations (cf. Nelson and Platnick 1984; Eldredge 1985, 2003). A taxon which contains all the descendents of its members as proper parts cannot be ancestral to any other such taxon. To see this point, consider the cladogram in Figure 7.1. If we ask what the ancestor of the highlighted clade A is, then the answer can only be a *species*, not another clade. Clade A is of course a *part* of the larger clade B, but it is not the *offspring* of B. For offspring must have an independent existence from their parents, and be able to outlive them.

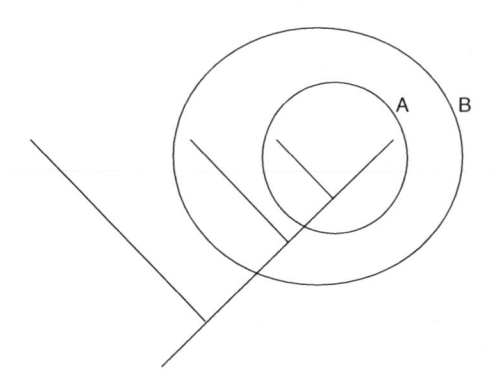

Figure 7.1. Cladogram showing two nested clades

[9] Though van Valen does not actually use the term 'clade selection', for he does not accept cladism.

But clade A cannot outlive clade B. The only way a monophyletic clade can cease to exist is if all of its constituent species go extinct, which implies that all the sub-clades which are parts of it must cease to exist too. If clade B ceases to exist, then clade A must do so too. So A is not the offspring of B; it is simply a part of it.[10]

It follows that Williams's idea that cladogenesis constitutes 'reproduction for clades' is incorrect. Reproduction means one entity giving rise to another entity of the same type, but clades cannot do this. In cladogenesis, the entity that splits is a species lineage. Presuming it splits into two (which is most usual), and given the standard cladistic convention that a species automatically goes extinct when it splits, the result is a new clade containing three species—the original one (now extinct), and two new ones. But the new clade is not the *offspring* of any of the clades to which the original species belonged, but rather a part of them. Williams's claim that there is no reason to focus on species selection rather than clade selection is therefore wrong. There is such a reason: species give rise to offspring, hence form ancestor–descendant lineages, but clades do not. Monophyletic clades are not the sorts of thing to which fitness can be meaningfully ascribed.

How might defenders of clade selection respond? One response would be to concede that clades do not reproduce, but argue that differential extinction of clades might still occur. This is true enough. However, selection on entities that do not reproduce their kind is not very interesting, and will not lead to adaptations. All sorts of entities are subject to selection in this weak sense. A collection of atoms may have different probabilities of radioactive decay, a collection of buildings may have different probabilities of being demolished, and so on. Natural selection is only an interesting idea when applied to entities that reproduce. Moreover, clade selection in this weak sense is not a more general version of species selection. It is precisely because species do reproduce that species selection is a potentially interesting evolutionary mechanism.

Note that this is *not* to say that interesting cases of natural selection must involve fecundity rather than viability selection, which is certainly untrue. Rather, the point is that natural selection, whether it operates

---

[10] In Okasha (2003a), this argument is expressed by saying that the parent–offspring relation must be capable of becoming the ancestor–descendant relation, i.e. if two entities are related as parent and offspring, it must be possible for them to *become* related as ancestor and descendant, at a later point in time. This means that death of the parental entity must not necessarily entail death of the offspring.

on differences in survival or fecundity, is only interesting when applied to entities that do *in fact* reproduce. Differential survival of organisms and species is interesting, because organisms and species reproduce their kind. Differential survival of clades is not, because clades do not.

A second possible response is to concede that clade fitness in the sense of number of offspring clades does not make sense, but to replace it with another notion. Why not let a clade's fitness refer to the number of sub-clades it comes to have as parts? And by heritability, we could mean resemblance between a larger clade and its sub-clades, rather than its offspring clades. Clade selection in this sense could help explain differences in 'bushiness' between clades. Fitter clades are the ones whose traits lead them to grow bushier than others.

One might object that redefining clade fitness this way means that clade selection ceases to be a genuinely Darwinian process. We do not normally appeal to differences in 'branch fitness' to explain why some branches of an oak tree contain more twigs than others, but structurally this is parallel. But in any case, there is a deeper objection to the proposed way of salvaging clade selection.

Consider the clades marked A and B in Figure 7.2, each containing two extant species. If clade A is fitter than clade B, according to the suggested redefinition, this means that A will come to contain more sub-clades as parts. But cladogenesis only occurs when species lineages split, so this means that the species in A must leave more offspring species than those in B. Therefore clade selection in the suggested sense is redundant—species selection can do all the work. The fact that clade A grows bushier than clade B is explained by the fact that the *species* in A are fitter than those in B. Defining clade fitness as number of sub-clades and then invoking a process of clade selection is pointless. For clade

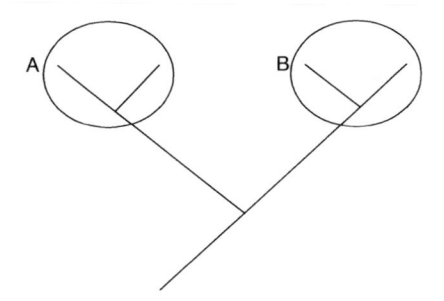

Figure 7.2. Cladogram showing two sister clades

selection in this sense explains nothing that is not already explained by species selection.

A third (related) response also argues for a redefinition of clade fitness. Why not define the fitness of a clade as the average fitness of its constituent species? In effect, this is to suggest that we should treat clade selection as an MLS1 process, in which the focal units are not the clades themselves but rather their constituent species. The relation between clades and species would thus be analogous to the relation between groups and individuals in most models of group selection. Clearly this would circumvent the problem that clades cannot reproduce their kind; for in MLS1 it is the particles not the collectives that must stand in parent–offspring relations.

Though logically coherent, it is hard to see what the *point* of treating clade selection this way would be. The point of a group selection model which defines group fitness as average individual fitness is to model situations where the fitness of an individual depends on the composition of its group. If there are no group-level effects on individual fitness, there is no need for a group selection model of this sort—individual selection is the only force at work. In the clade case, there are unlikely to be clade-level effects on species fitness, in the way there are often group-level effects on individual fitness. Why should the fitness of any species depend on which other species are in its clade? Such a dependence is of course possible. For example, if a given species goes extinct, this could increase the fitness of a sister species with which it competes for resources. But it is unlikely to be a common phenomenon.

Therefore, treating clade selection as an MLS1 process, with species as the focal units, is conceptually coherent but unlikely to have useful empirical application. If average species fitness is greater in clade A than clade B, then clade A will become bushier than B. But almost certainly, this falls within the purview of *species* selection—the clade-level trend is a by-product. Nothing is gained by defining clade fitness as average species fitness and then attributing the difference in bushiness to clade selection. In the absence of systematic clade-level effects on species fitness, this is artificially to multiply levels of selection for no reason.

The foregoing arguments suggest that the concept of clade selection is at worst incoherent and at best simply collapses into species selection. I suggest that evolutionists should therefore abandon the concept.[11]

---

[11] Though see Haber and Hamilton (forthcoming) for an attempt to salvage clade selection from the objections presented here.

Finally, let us consider a different way in which clades might be relevant to species selection. It is sometimes suggested that species selection can only take place among the members of a monophyletic clade; Vrba (1989) makes this part of her definition of species selection. Note that this has nothing to do with treating clades themselves as units of selection; rather, the suggestion is that the clade is the entity whose composition gets changed by species selection, that is, the analogue of what in ordinary Darwinian selection is usually called the 'population'.

I think that this suggestion unnecessarily restricts the concept of species selection (cf. Grantham 1995). In general, the population of entities undergoing natural selection, at whatever level, is demarcated by the requirement that its members experience a common selective force (cf. Sober and Lewontin 1982). This requirement was implicit in the abstract treatment of selection developed in previous chapters; for it is implied by the idea that character differences must cause fitness differences, within the population of entities undergoing selection. Nothing follows about the genealogical connections, if any, among the entities in the population.[12]

This point holds good for species selection too. It *may* be the case that the species in a monophyletic clade are more likely to be subject to common causal influences than those in a non-monophyletic group. For selective forces are often phylogenetically mediated, as Sterelny (1996a) notes. Just as conspecific organisms are often affected similarly by environmental changes, the same may be true of con-cladistic species. But this is an empirical issue. It does not justify making monophyly a condition of any set of species whose composition can be changed by species selection. Again, this point is illustrated by the hypothesis that species selection favoured sexual reproduction. As noted in Section 7.3, if this hypothesis is correct, then the set of species on which selection acted consists of all those species, sexual and asexual, whose fitness was affected by their capacity to evolve. There is no a priori reason to think that this set was monophyletic.

---

[12] Grantham (1995) argues that 'the term population has a genealogical component ... and an ecological component' (p. 311), on the grounds that populations are generally assumed to contain conspecific organisms. This is admittedly how the term 'population' is often used. But the requirement of conspecificity is indefensible. Just as selection can act on a population of asexual organisms, so it can act on a population of sexual organisms which do not exchange genetic material with each other.

8

Levels of Selection and the Major Evolutionary Transitions

INTRODUCTION

In Chapter 1, we noted that the levels-of-selection question has undergone a subtle transformation in recent years. In early discussions, the existence of hierarchical organization was taken for granted, as if it were simply a brute fact about the biological world; the aim was to understand selection and adaptation at pre-existing hierarchical levels. But recent theorists, beginning with Buss (1987), have sought to understand how the biological hierarchy evolved in the first place, thus transforming the levels-of-selection question. This chapter examines the implications of this change in focus.

Maynard Smith and Szathmáry (1995) discuss the evolution of hierarchical organization under the heading 'the major transitions' in evolution; in a similar vein, Michod (2005) talks about 'evolutionary transitions in individuality'. These transitions include: (i) solitary replicators → networks of replicators enclosed in compartments; (ii) unlinked genes → chromosomes; (iii) prokaryotic cells → eukaryotic cells with intra-cellular organelles; (iv) single-celled → multicelled organisms; and (v) solitary organisms → colonies. In each case, a number of smaller units, originally capable of surviving and reproducing on their own, formed themselves into a larger unit, creating a new level of organization. As Maynard Smith and Szathmáry say, 'entities that were capable of independent replication before the transition can replicate only as part of a larger whole after it' (1995 p. 6). The challenge is to understand these transitions in Darwinian terms. Why was it advantageous for the lower-level units to sacrifice their individuality and form themselves into a corporate body? And how could such an arrangement, once first evolved, be evolutionarily stable?

Levels of Selection and the Major Evolutionary Transitions 219

This immediately raises the levels-of-selection issue. For during an evolutionary transition, lower-level selection may frustrate the evolution of the higher-level unit. In the transition to multicellularity, for example, selection between competing cell lineages may disrupt the integrity of the emerging multicelled creature. However, if selection also acts on the higher-level units, this may promote the evolution of adaptations for suppressing internal conflict. Thus in the multicellularity case, Buss (1987) argues that germ-line sequestration is such an adaptation, for it reduces the probability that mutant cells, arising during ontogeny, will pass to the next generation. This particular argument has been contested, but the general idea that evolutionary transitions involve an interaction between levels of selection is widely accepted. In large part, this explains the recent resurgence of interest in multi-level selection theory among biologists (cf. Reeve and Keller 1999).

Section 8.1 explores how the traditional levels-of-selection question has been transformed by the recent developments. Section 8.2 looks at competing methodologies for studying evolutionary transitions; the 'genic' approach of Maynard Smith and Szathmáry (1995) is contrasted with the 'hierarchical' approach of Buss (1987). Section 8.3 asks what becomes of the contrast between MLS1 and MLS2 in the context of the major transitions; I argue that both types of multi-level selection are relevant, but at different stages of a transition. This point is illustrated in Section 8.4, with reference to Michod and co-workers' models of the transition to multicellularity (Michod 1997, 1999, 2005; Michod and Roze 1999; Roze and Michod 2001; Michod and Nedelcu 2003). Section 8.5 draws some tentative conclusions.

8.1 THE TRANSFORMATION OF THE LEVELS-OF-SELECTION QUESTION

Traditionally the levels-of-selection question has been formulated in a 'synchronic' way. The biological hierarchy, with its various levels of nestedness, is assumed to be in place; the question is about the level(s) at which selection currently acts, or acted in the recent past. But as Griesemer (2000) notes, such formulations say nothing about how the biota came to be hierarchically organized in the first place. By contrast, in recent work on evolutionary transitions, the levels question has been formulated diachronically, as a question about how the biological hierarchy originally evolved. From the diachronic perspective, the traditional approach is not wrong; it is just incomplete.

Moving from a synchronic to a diachronic construal of the levels question is an instance of a more general trend in evolutionary theorizing, namely 'endogenizing' parameters that were once treated as exogenous. Consider for example the mutation rate. Traditional genetical models treat the mutation rate as exogenous—a parameter that affects the outcome of evolution, but does not itself receive an evolutionary explanation. But this approach is not wholly satisfactory, as has long been recognized (cf. Maynard Smith 1966). The mutation rate varies widely within and between taxa, and is affected by complex molecular machinery, so must itself have evolved. A complete evolutionary theory must therefore try to endogenize the mutation rate, not treat it as a given. Another example is fair Mendelian segregation. Early evolutionists assumed that segregation obeys the Mendelian law and deduced the consequences. But there is an important question about *why* segregation is usually Mendelian, that is, how it evolved to be that way. Again, parameters that were once treated as exogenous must themselves be given an evolutionary explanation.

The importance of not helping ourselves to things that require explanation is emphasized by Buss (1987), who argues that 'individuality' is a derived not an ancestral character (p. 25). By 'individuality' he means features such as genetic homogeneity, early germ-line sequestration, and cellular differentiation, which characterize modern metazoans.[1] Buss's point is not just that metazoans have not always existed; this much is obvious. Rather his point is that neo-Darwinism, while purporting to be a general theory of evolution, has incorporated assumptions which restrict its scope and thus compromise its generality. That organisms are largely genetically homogenous, that their parts work for the good of the whole, that within-organism selection is of no evolutionary consequence, and that acquired characters cannot be inherited are among the basal assumptions of neo-Darwinism. Buss observes firstly that these assumptions are not true of all taxa; and secondly, to the extent that they are true, they themselves require evolutionary explanation.[2] In effect,

---

[1] Buss uses the term 'individual' to mean a multicelled organism; by contrast, many theorists concerned with the major transitions use 'individual' to mean any entity which functions as a unit of selection.

[2] Thus Buss writes: 'a theory which assumes individuality as a basal assumption cannot be expected to explain how individuality evolved' (1987 p. 25).

the work of Maynard Smith and Szathmáry (1995) involves extending Buss's approach to all levels of hierarchical organization.[3]

How significant is the shift from a synchronic to a diachronic formulation of the levels-of-selection question? It might be argued that the shift is less dramatic than I have implied. After all, participants in the early discussions clearly knew that entities at different hierarchical levels form a temporal sequence, 'below' corresponding to 'evolved before'; no one thought that the earliest life forms were hierarchically complex. And much of the early sociobiological literature, for example, E.O. Wilson (1975), was an attempt to explain how eusocial insect colonies evolved—though the language of 'evolutionary transitions' was rarely used. Despite these points, there are three respects in which the traditional discussions are theoretically inadequate for understanding evolutionary transitions.

The first concerns the very concepts used to understand natural selection, a point alluded to in Chapter 1. Consider for example Dawkins's concept of a replicator. Dawkins characterizes replicators in terms of high-fidelity copying, which enables 'genetic information' to be faithfully transmitted across generations. But the molecular mechanisms that permit this are evolved features; and arguably, the notion of genetic information only makes sense thanks to the genetic code, which is also the product of evolution (Griesemer 2000; Godfrey-Smith 2000b). These are not objections to the replicator concept per se, but they do suggest that it cannot be the basis for a wholly general theory of evolution. Similar objections apply to Dawkins's concept of a vehicle. Dawkins (1982) requires that vehicles exhibit a high degree of internal cohesion—a requirement which rules groups out but organisms in, he says (p. 115). But again, this puts the cart before the horse. As Michod (1997) stresses, organisms are themselves groups of cooperating cells; their cohesion is the result of adaptations that suppress within-group competition. If we wish to understand show cohesiveness evolved initially, we clearly cannot build it into the concepts used to formulate evolutionary theory.

Secondly, traditional discussions often privileged multicelled organisms over all other entities, despite paying lip-service to hierarchical organization. Symptomatic of this was the use of 'individual' to mean

---

[3] Maynard Smith and Szathmáry (1995) note their indebtedness to Buss, while disagreeing with a number of his specific arguments concerning the evolution of germ-line sequestration and maternal control (p. 244–6).

multicelled organism, and the assumption that 'individual selection' is all pervasive while 'group selection' is a theoretical curiosity. From an evolutionary transitions perspective, this is clearly unsatisfactory. Higher-level selection invariably occurs during an evolutionary transition, for the higher-level unit, or group, has got to suppress competition among its parts in order to emerge as a genuine whole, with adaptations of its own. (In Michod's terms, fitness must be 'exported' from the lower to the higher level.) Moreover, we should not think of 'individual' and 'group' as denoting absolute levels in the hierarchy, as traditionally assumed; they are strictly relative designations. An entity that is an individual in one context may be a group in another, and vice versa. Queller (2000) makes precisely this point when he argues that insect colonies, to the extent that they function as adaptive wholes, should be regarded not as superorganisms but simply as *organisms*. This may sound odd; but it captures the idea that no absolute meaning attaches to levels in the hierarchy—just as no absolute meaning attaches to traditional taxonomic ranks, according to cladists.[4]

Thirdly, the traditional discussions often failed to appreciate the thematic similarities between selection at different levels. Consider for example Dawkins's (1982) discussion of how the first replicators may have become compartmentalized. Dawkins says that it is 'easily understood' why independent replicators might have gained an advantage by 'ganging up together' into cell-like compartments, because their biochemical effects might have complemented each other (p. 252). This is a plausible idea; see the discussion of hypercycles below. But what Dawkins misses is that it in effect invokes group selection. From the selective point of view, replicating molecules combining themselves into compartments is strictly analogous to individual organisms combining themselves into colonies or groups (Szathmáry and Demeter 1987). But Dawkins is an implacable *opponent* of group selection, insisting on the impotence of selection for group advantage as an evolutionary mechanism. Clearly, Dawkins does not appreciate that evolutionary transitions necessarily involve selection at multiple levels.

The thematic similarities between the various transitions are emphasized by Michod (1999) and Maynard Smith and Szathmáry (1995). Cooperation among lower-level units and suppression of within-group competition are important in all the transitions—without them, no

---

[4] This idea is the basis of the 'rank free' phylogenetic systematics advocated in the *Phylocode* project; see <http://www.ohiou.edu/phylocode/>.

higher-level units can evolve. Mechanisms that promote cooperation include kinship, population structure, synergistic interactions, and reciprocation; mechanisms that suppress competition include division of labour, randomization (e.g. fair meiosis), policing by fellow group members, and vertical transmission. Many of these themes were originally developed in the sociobiological literature of the 1970s and 1980s, often in relation to social insects and/or animal behaviour, but their applicability is far more general. Thus Michod (1999) says that the 'set of tools and concepts' developed by sociobiologists to study social insects 'has proved useful for studying the other major transitions' (p. 8); Queller (1997) argues similarly.

Eigen and Schuster's well-known work on molecular hypercycles illustrates this point well (Eigen and Schuster 1977; Eigen 1979). Their theory tries to explain why independent RNA molecules, capable of surviving and reproducing alone, might have formed cooperative networks or communities; this was probably one of the earliest evolutionary transitions. Since RNA can perform enzymatic functions, that is, can catalyse the replication of other RNAs, cooperative interactions between different replicators could easily have arisen, if their catalytic effects were complementary.[5] Figure 8.1 depicts a three-membered hypercycle, where each member is an RNA sequence of a different type. Type A catalyses the replication of type B, which in turn catalyses the replication of type C, which in turn catalyses the replication of type A. (Note that catalytic effects exhibit specificity, i.e. the As only help replicate Bs, Bs only help replicate Cs, and so on; this is an essential feature of hypercycles.) It is clear that hypercycles can evolve by natural selection—for any individual molecule within a hypercycle will replicate faster than a free-living molecule of the same type, thanks to the fact that the circle closes. In theory, this may explain how networks of replicating molecules first arose.

As Michod (1999) emphasizes, hypercycles constitute rudimentary cooperative groups; so the familiar themes of cooperation versus conflict, individual versus group interests, selfish versus altruistic mutations, and so on apply to them. In this context, selfish mutations are ones that make an RNA molecule a worse replicase but a better target, so are individually beneficial but harmful for the whole hypercycle; altruistic mutations do

[5] Eigen's original work assumed that the RNA replicators coded for protein replicases, and thus that translation machinery had evolved; but this is an inessential feature, as Maynard Smith and Szathmáry (1995) note, given that RNA can itself perform enzymatic functions (p. 52).

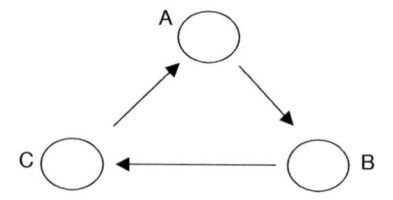

Figure 8.1. Three-membered hypercycle

the converse. Within a free-mixing or homogenous population, altruism cannot easily spread. The reason for this was discussed in Chapter 5 — the evolution of (strong) altruism requires positive association between altruists. This moral applies to prebiotic evolution just as much as to the evolution of animal behaviour. So it is no surprise that the next step is to consider how RNA molecules might have become compartmentalized, permitting evolutionary outcomes not possible in an unstructured population (Szathmáry and Demeter 1987). This idea is discussed below; for the moment, the point is just that evolutionary transitions occurring at quite different hierarchical levels are thematically analogous.

Most philosophical discussions of the levels of selection have adopted a synchronic perspective; though Griesemer (2000) and Michod (1999) are exceptions. From a diachronic perspective, many of the philosophical issues assume a different aspect. Consider, for example, the realism/pluralism issue. In one respect, pluralism is less plausible from a diachronic perspective. If eukaryotic cells do not exist, for example, then it is an objective fact that selection does not operate at the eukaryotic cell level. So from a diachronic perspective, it is obviously untrue that the choice of level(s) of selection is *entirely* unconstrained by reality, as certain pluralists have maintained. But in another respect, pluralism is more plausible when the levels question is understood diachronically. For during evolutionary transitions, borderline cases of part–whole structure are inevitable, and such cases provide a natural home for pluralism, as discussed in Chapter 4. If it is indeterminate whether entities at level X exist, then it may be correspondingly indeterminate whether selection acts at level X. So the realism/pluralism debate looks significantly different from a diachronic perspective.

The same is true of reductionism in one of the senses distinguished in Chapter 4, namely the idea that higher-level phenomena should be explained from the 'bottom up', rather than at their own level. This choice between reductionistic and holistic explanatory strategies arises wherever there is part–whole structure, not just in biology, and may be influenced by a variety of factors. But if what we want to explain is the evolution of hierarchical structure itself, the problem assumes a unique aspect. In particular, a reductionist strategy seems mandatory. Given that collectives were formed by lower-level particles coming together to form a whole, explanation 'from below' is surely essential, at least in the early stages of a transition; for we need to understand the selective forces that led the particles to abandon their free-living existence. So although some collective-level phenomena might usefully be explained 'at their own level', the original evolution of the collectives is not one of them. We shall return to this issue below.

8.2 GENIC VERSUS HIERARCHICAL APPROACHES TO THE TRANSITIONS

In broad terms, it is clear what happened in evolutionary transitions—lower-level units coalesced into larger ones—though the precise steps involved cannot be known with any certainty. It is also clear that a Darwinian approach to the transitions is essential—we need to understand the selective forces at work, not just the mechanistic details of how the coalescing happened. (For example, to understand the origin of the eukaryotic cell, we need to know why, not just how, ancient prokaryotic cells came to contain organelles, i.e. what were the adaptive advantages, and for whom.) Despite agreement on these basic points, not all theorists agree on the correct methodology for studying evolutionary transitions.

In *The Evolution of Individuality*, Buss makes some interesting methodological remarks about how the transitions should be studied. He contrasts the 'hierarchical' approach, which he himself favours, with an alternative 'genic' approach. According to Buss, the former has its roots in the work of Roux and Weismann, both of whom recognized a multiplicity of units of selection, while the latter stems from the work of G.C. Williams and Dawkins. Buss describes the choice between the two approaches in conventionalist terms—it is a choice of language, not empirical fact, he claims (p. 54). However, he regards the genic approach as inferior for two reasons. First, it simply records the outcome of evolution without providing causal explanations; this is a version of the 'bookkeeping objection' discussed in Chapter 5. Secondly, it ignores the complex cross-level interactions that drive evolutionary transitions. Buss writes: 'to concentrate solely on the selfishness of each evolutionary innovation is to miss what might be learned from the study of potential synergisms and conflicts between units. To adopt a gene selection perspective is not wrong. It simply does not help unravel the central dilemma of our science' (p. 55).

Falk and Sarkar (1992) argue that Buss does the hierarchical approach a disservice by suggesting that the choice between it and the genic approach is merely conventional. The hierarchical perspective, they say, 'permits more than a different interpretation of data already available from the genic selectionist perspective' (p. 468). They point out that according to Buss's own theory, various features of metazoan ontogeny, for example, germ-line sequestration and maternal control of early development, arose as a result of conflict between selection at the cellular and organismic levels. Whether this is true is presumably a matter of objective fact; indeed, there are rival theories for why germ-line sequestration and maternal control evolved.[6] Given that factual issues of this sort are at stake, Buss's conventionalism is misplaced, Falk and Sarkar argue. This illustrates the point made above, that pluralism about levels of selection looks odd from a diachronic perspective.

It is clearly true that hypotheses about how evolutionary transitions occurred are factual not conventional, as Falk and Sarkar say; and formulating such hypotheses is most natural using a hierarchical or multi-level selection framework. However, as we saw in Chapter 5, selection processes that occur at different hierarchical levels can often be usefully viewed from a gene's-eye perspective, for they often eventuate in gene frequency change. So if Buss's point is just that the gene's-eye perspective does not constitute an *empirical* alternative to multi-level selection, then to this extent his conventionalism is correct.

Maynard and Szathmáry (1995) argue that a genic approach *is* the best way to understand the major transitions, contra Buss. They say: 'the transitions must be explained in terms of immediate selective advantage to individual replicators: we are committed to the gene-centred approach outlined by Williams (1966), and made still more explicit by Dawkins (1976)' (p. 8). Using selfish gene reasoning, Maynard Smith and Szathmáry show that some of Buss's hypotheses about the selective forces that drove the transition to multicellularity are unlikely to be correct. (In a similar vein, many of Buss's critics observed that he ignores

---

[6] See Maynard Smith and Szathmáry (1995) p. 245 for criticism of Buss's view; see Michod (1999) ch. 6 for extended discussion, in a quantitative framework, of the possible adaptive advantages of germ-line sequestration.

the most fundamental reason why the cells in an organism usually cooperate: they are genetically identical.) Despite this, Maynard Smith and Szathmáry's approach is simultaneously hierarchical. They agree that higher-level units evolved by the coalescing of lower-level units, and they agree that conflict and synergy between units played a critical role. Furthermore, they accept Buss's argument that the integrity of the higher-level unit requires adaptations for suppressing conflict among the parts. Although Maynard Smith and Szathmáry rarely use the language of hierarchy and multi-level selection, much of their discussion could easily be reformulated in such terms.

What does this show? I think it shows that Queller (1997) is correct when he argues that we do not need to choose between the hierarchical and genic approaches—'we can and must have it both ways' (p. 187). Buss's view to the contrary stems from underestimating the genic approach. If the genic approach consisted of no more than saying that each evolutionary transition occurred because it benefited selfish genes, then Buss's objection would be well taken. (As Buss rightly notes, Dawkins's explanation for why replicators 'packaged' themselves into organisms—'the world...tends to become populated by mutually compatible sets of successful replicators, replicators that get on well together'—explains very little (quoted by Buss 1987 p. 180).) However, Maynard Smith and Szathmáry advance detailed hypotheses about the actual selective forces at work in each of the transitions, so they clearly cannot be accused of 'mere bookkeeping'. They show that thinking in terms of genetic self-interest can be valuable even when we are trying to explain the evolution of hierarchical organization itself; but their explanations do *not* degenerate into the post hoc descriptions of evolutionary change that some gene's-eye theorists have been guilty of. This confirms the view defended in Chapter 5, that the genic and multi-level approaches are complementary, not antithetical.

Michod's work on evolutionary transitions nicely illustrates this complementarity; his work constitutes a methodological middle-ground between Buss and Maynard Smith and Szathmáry. Like Buss, Michod uses hierarchical language and conceptualizes selection in multi-level terms. The main task, as he describes it, is to explain how higher-level entities became 'Darwinian individuals', that is, how they acquired the properties of heritable variation in fitness (Michod 2005). Cooperative interactions between lower-level units were the first stage, Michod argues; they provided the context in which fitness could be 'exported' from the lower to the higher level. However, a group of cooperating units only possesses true 'individuality' once it evolves adaptations for suppressing within-group conflict. Michod's account of how such adaptations evolve arises directly out of multi-level selection theory. Conflict-suppressors reduce the potential for within-group change, which increases heritability at the group level, and enhances the power of between-group selection relative to within-group selection. In this way the group eventually becomes a cohesive, harmonious unit. Unsurprisingly, Michod's description of this process employs the Price formalism.

However, Michod's approach is simultaneously 'genic', for he uses formal population-genetic models. His basic model for the evolution of multicellularity employs standard two-locus modifier techniques to explore the spread of cooperation between cells in the emerging multicelled organism, and the subsequent evolution of conflict-reducing adaptations (Michod 1997, 1999; Roze and Michod 2001). The first locus determines whether a cell cooperates (C) or defects (D); cooperators perform somatic functions that benefit the whole organism, while defectors abandon somatic duties for faster replication. So within-organism selection favours the C allele, while between-organism selection favours D. Which level of selection dominates depends on parameters such as the mutation rate, the rate of cell division, adult size, and others. The second locus affects these parameters, thus potentially altering the balance between levels of selection. For example, an allele at the second locus might reduce development time and thus adult size, or reduce the rate of somatic mutation.[7] If this allele can spread at its own locus, it will help promote the cohesiveness of the organism, by reducing the strength of within-organism selection for the D allele at the first locus.

This brief description of Michod's model shows how the genic and hierarchical approaches can peacefully coexist. The idea of selection at more than one hierarchical level, and the themes of cooperation, conflict, and synergistic interaction between units, emphasized by Buss, are integrated with standard population-genetic analysis by Michod. So Michod's model conforms to Maynard Smith and Szathmáry's dictum that 'the transitions must be explained in terms of immediate selective advantage to individual replicators', while at the same time lending itself naturally to a description in hierarchical language, which Maynard Smith and Szathmáry themselves eschew. (Symptomatic of this is that the criticism levelled against Buss's original theory—ignoring the significance of

---

[7] Somatic mutation does affect the between-generation gene frequency change, as Michod is assuming that germ-soma differentiation has not yet evolved.

genetic relatedness between cells—clearly does *not* apply to Michod's theory, in which kinship plays a central role.) No clearer illustration of the compatibility of the two approaches could be hoped for.

8.3 MLS1 VERSUS MLS2 IN RELATION TO EVOLUTIONARY TRANSITIONS

Multi-level selection plays a key role in theorizing about evolutionary transitions, as we have seen. This raises an important question. Is the relevant type of multi-level selection MLS1 or MLS2? Recall that in MLS1, particles are the focal units and collective fitness is defined as average particle fitness; in MLS2, particles and collectives are both focal units and their fitnesses are independently defined. Damuth and Heisler (1988) originally formulated the MLS1/MLS2 distinction with reference to a synchronic formulation of the levels question, that is, hierarchical structure was assumed to be already in place. But what becomes of the distinction from a diachronic perspective? If we wish to use multi-level theory to explain evolutionary transitions, is MLS1 or MLS2 better suited to the job?

A natural first response is to say that MLS2 is the relevant type of multi-level selection, in the context of evolutionary transitions. For MLS1 explains the evolution of particle characters, not collective characters, as discussed in Chapter 2. In an evolutionary transition, new collectives come into existence from an ancestral state in which they did not exist. On the standard view, this requires the collectives to evolve adaptations for reducing conflict between their constituent particles, for example, germ-line sequestration. These adaptations are properties of the collectives themselves, not their constituent particles. So the collectives themselves are surely the focal units. Therefore, collective-level selection must mean selection between collectives based on differential production of offspring *collectives*, not particles.

This argument sounds right, and in part it is. But it is not the whole story, for it only applies to the later stages of an evolutionary transition, when the collectives have already evolved as discrete units, with life cycles of their own, hence capable of having fitnesses in the MLS2 sense. In the early transitional stages, matters are different. These early stages are characterized by the spread of cooperation among the particles, a prelude to the day when they will sacrifice their individuality entirely and form discrete collectives. (As Michod notes, conflict-reducing adaptations, which 'legitimize the new unit once and for all', come *after* the initial spread of cooperation (1999 p. 42).) So we should expect traditional theories for the evolution of cooperation/altruism, for example, kin and group selection, to be relevant to the early stages of an evolutionary transition. And as we saw in Chapter 4, kin and group selection models are usually of the MLS1 type: they are concerned with the spread of an *individual* phenotype, for example, altruism, in a group-structured population, and thus define group fitness as average individual fitness. So in the early stages of a transition, before cohesive collectives have evolved, multi-level selection of the MLS1 type is relevant.

If this is right, it suggests that both types of multi-level selection may be relevant to the major transitions, each at a different stage. An examination of the biological literature confirms this argument.

Recall the hypercycle concept. Hypercycles were proposed as an attempt to explain how RNA molecules might have formed cooperative networks, but they are susceptible to breakdown by selfish mutants. How can this problem be solved? Michod (1983, 1999) suggests that D.S. Wilson's 'trait-group' model may provide the key; a similar idea is explored by Szathmáry and Demeter (1987) and Maynard Smith and Szathmáry (1995). The trait-group model turns on the fact that population structure permits evolutionary outcomes not possible in an unstructured, or freely mixing, population. As we saw in Chapter 6, if trait groups are formed randomly, then weak altruism can evolve; if they are formed assortatively, then strong altruism can evolve. In the prebiotic context, altruists are RNA molecules that code for replicase, which boosts the replication rate of those around them. If RNA molecules on a surface become clustered into groups, then groups containing lots of replicase-coders may out-reproduce groups containing fewer, even though within each group the selfish types have an advantage. (Since neighbouring RNAs are likely to be relatives, kinship can easily lead to positive assortment.) Eventually, molecules may become compartmentalized into proto-cells, enclosed by a cell membrane, permitting even stronger selection for group advantage.

There are two crucial points to note here. First, the 'trait groups' of RNA molecules, which initially consist of independent molecules engaged in fitness-affecting interactions, are the precursors of the compartmentalized proto-cells that eventually evolve. Secondly, the type of group selection involved is MLS1. Groups containing lots of altruists 'out-reproduce' other groups in that they contribute more *individuals* to the next generation. So what the trait-group model explains is the spread of an *individual* trait—altruism—in the overall population, not the spread of a group trait. This is true both of Wilson's original trait-group model and of its applications to prebiotic evolution by Michod, Maynard Smith, Szathmáry, and Demeter. (As Maynard Smith and Szathmáry say, the aim of the model is to explain 'an increase in the frequency of altruistic replicators' (1995 p. 55).) This demonstrates that in the early stages of an evolutionary transition, multi-level selection of the MLS1 type is relevant. For although MLS1 can only explain the evolution of particle traits, not collective traits, early transitional stages are precisely characterized by the spread of a particle trait—cooperation. For groups of cooperating particles represent the first stage in the transition to a new collective.

Michod's work on multicellularity illustrates how MLS2 becomes relevant later in a transition. Recall the life cycle that Michod's model assumes (cf. Chapter 3, Section 3.5.1). A multicelled organism or proto-organism begins life as a propagule containing N cells, all of which come from the same parent. (If $N=1$, then the life cycle starts from a single-celled stage; if $N>1$ then reproduction is vegetative.) Cells are of two types, cooperators (C) and defectors (D). The propagule develops and grows by cell division, reaches adulthood, and then reproduces by emitting propagules (see Figure 8.2). There are two levels of selection: between cells within organisms, owing to different cell-types dividing at different rates, and between organisms, owing to different organism-types producing different numbers of propagules. Cell and organism selection thus operate over different timescales. However, the two levels of selection interact. Cellular selection reduces the fidelity with which organisms transmit their characters to their offspring, and thus contributes to transmission bias, hence reduced heritability, at the collective level. (For example, an organism that starts life as a three-celled CDD propagule may, as a result of cellular selection, contain mainly D cells as an adult, and thus give rise to propagules containing mainly D.) The overall evolutionary dynamics depend on both levels of selection.

This is an MLS2 model, because organisms and cells are both focal units, and their fitnesses are independently defined. As Michod and Roze (1999) note, 'the fitness of the organism, or cell-group...is the absolute number of offspring groups produced...[while]...the fitness of the cell is defined in terms of its [within-organism] rate of replication and survival' (p. 57). So an organism's fitness is *not* equal to the mean fitness of the cells within it, although these two quantities may be proportional to each other. This is the defining mark of an MLS2 theory.

232   Evolution and the Levels of Selection

Propagules   Adult organisms   Offspring propagules

O - defector   ● - cooperator

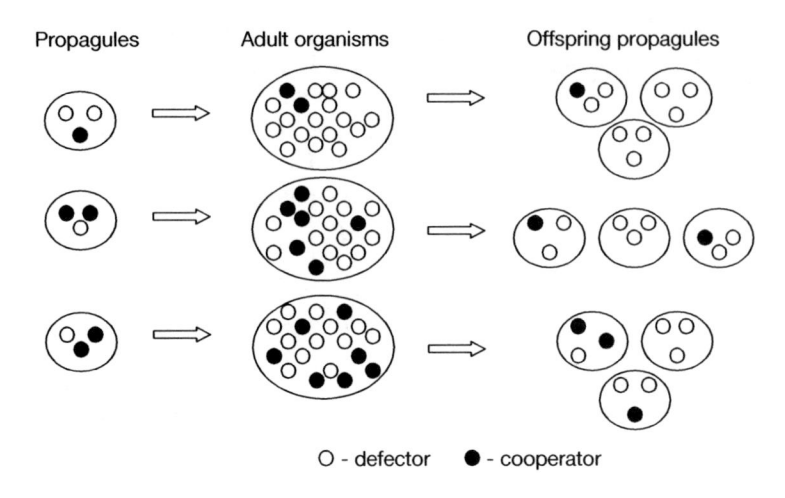

Figure 8.2. Michod's model for evolution of multicellularity

Michod and Nedelcu (2003) make some interesting remarks about the relation between the fitnesses at the two levels, in this model. They argue that during an evolutionary transition, the fitness of the higher-level unit must be 'decoupled' from the fitness of the lower-level units of which it is composed: 'group fitness, is, initially, taken to be the average of the lower-level individual fitnesses; but as the evolutionary transition proceeds, group fitness becomes decoupled from the fitness of its lower-level components' (p. 66). This remark bears directly on the MLS1 versus MLS2 issue. For in effect, Michod and Nedelcu are saying that in the early stages of a transition, collective fitness is defined in the MLS1 way, but as the transition proceeds and fitness decoupling occurs, collective fitness in the MLS2 sense becomes relevant. The intuitive grounds for this view are clear. In the early stages of a transition, where cooperative interactions among the particles are just beginning to spread, discrete collectives with life cycles of their own do not exist, so the notion of collective fitness as number of offspring *collectives* cannot apply. Later in the transition this notion can apply, so the salient type of multi-level selection becomes MLS2 (cf. Okasha 2006; Michod 2005).

What exactly does it mean for collective fitness to be 'decoupled' from particle fitness? This is less simple than it appears at first sight. For it does not *just* mean defining collective fitness in the MLS2 rather

Levels of Selection and the Major Evolutionary Transitions   233

than the MLS1 way, though this is part of it. To explore this question further, we need to look at Michod's model in more detail.

8.4 MICHOD ON FITNESS DECOUPLING AND THE EMERGENCE OF INDIVIDUALITY

Consider again the model life cycle depicted in Figure 8.2. This model presumes that the transition has proceeded far enough so that the multicelled organisms can possess fitnesses in the MLS2 sense, that is, a process of reproduction at the organismic level takes place. An organism's fitness is its number of offspring propagules; this number may be influenced by various factors. One important factor, Michod argues, is the frequency of cooperating (C) cells in the adult. The higher this frequency, the greater the functionality of the adult organism, so the more propagules it will produce. Therefore, Michod and Roze (1999) suggest the following expression for the fitness of the ith organism:

$$W_i = 1 + \beta q_i' \quad (8.1)$$

where q_i' is the frequency of C cells in the adult, and β is a parameter measuring the degree to which cooperation among cells benefits the organism.

Is expression (8.1) satisfactory? Clearly, this depends on whether frequency of C cells in the adult is the *only* factor that affects an organism's propagule production. Michod and Roze argue that another factor may be *adult size*, that is, total number of cells in the adult. Since there is no separate germ-line, the greater the number of cells in the adult, the more offspring propagules it can send out. Also important is the size of the propagules themselves—the smaller they are, the more of them can be produced. Therefore, Michod and Roze consider an alternative expression for organismic fitness:

$$W_i = (1 + \beta q_i') k_i / N \quad (8.2)$$

where k_i is the total number of cells in the adult, and N is propagule size. So expression (8.2) captures the idea that organismic fitness depends upon both adult functionality *and* adult size. Importantly, expression (8.2) does not imply that selection will *automatically* favour lower values of N, because as Michod and Roze note, adult size k_i may itself depend linearly on N, in which case N will cancel when k_i is expanded (1999 p. 10). Clearly, k_i will depend positively on N, that is, larger propagules will turn into larger adults, but the dependence may or may not be linear. If the dependence is linear, this means there is no *intrinsic* advantage to producing lots of small propagules rather than a few large ones, because the large ones will grow to be a larger size, hence be fitter themselves.[8]

Note that expressions (8.1) and (8.2) are not competing *definitions* of organismic fitness. On the contrary, both define organismic fitness identically, as number of offspring propagules. Rather, they make competing claims about what organismic fitness, so defined, depends on. Expression (8.1) says that organismic fitness depends on adult functionality alone; (8.2) says that it also depends on adult size. So they represent alternative modelling assumptions, not alternative definitions.

Michod and Roze argue that the choice between (8.1) and (8.2) raises 'interesting issues' concerning the extent to which the emerging multi-celled creatures possess 'individuality' (p. 56). For fully fledged organisms such as ourselves, the number of cells we contain as adults does not directly affect our fitness—there is no particular advantage to being fatter. However, Michod and Roze argue that for creatures 'on the threshold of multicellular life', fitness *will* probably depend on adult size, so is better captured by expression (8.2). However, this means that the creatures lack true 'individuality', because there is 'a direct contribution of cell fitness to organism fitness' (p. 10). This is because adult size is *itself* dependent on cell fitness—an organism whose constituent cells are very fit, that is, divide very fast, will achieve a larger adult size. So if organismic fitness is given by expression (8.2) rather than (8.1), then differences in organismic fitness may stem directly from differences in cellular fitness. In Michod and Roze's terms, this means that organismic fitness has not been 'decoupled from the fitness of the component cells', which in turn reflects the fact that the organism has not evolved true individuality (p. 57).

This is important for three reasons. First, it helps elucidate the concept of 'fitness decoupling'. For fitness at the two levels to be decoupled, it is *not* sufficient that organismic fitness be defined in the MLS2 rather than the MLS1 way. What is needed, in addition, is that differences

---

[8] Linear dependence of k_i on N has another interesting consequence. It means that an organism's fitness in the MLS2 sense—given by expression (8.2)—will be directly proportional to its fitness in the MLS1 sense, i.e. to the total number of offspring *cells*, rather than propagules, that it produces; see Chapter 2, Section 2.2.3 for further discussion of this point.

in organismic fitness should not arise solely through differences in cell fitness. Michod and Roze (1999) give a nice example of how this condition can fail to be satisfied. Suppose that there are no interactions between cells, that is, no cooperation or defection, but that different cell types have different intrinsic rates of replication. Suppose C cells divide faster than D cells. Therefore, an organism starting life as a CCC propagule will achieve a larger adult size than one starting life as a DDD propagule, so will have more offspring. (In terms of expression (2), the first organism has a higher value of k_i/N than the second.) This means that differences in organismic fitness are side effects of differences in cell division rates, and have nothing to do with organismic functionality; so fitness decoupling has not been achieved. Indeed in a sense these 'organisms' are not worthy of the name. Only once adult functionality, rather than adult size, becomes the main determinant of fitness does the emerging cell-group constitute a proper organism.

Secondly, Michod and Roze's discussion brings out an interesting link between fitness decoupling and the concept of a cross-level by-product, developed previously. (The example of C and D cells with different intrinsic rates of replication is reminiscent of Sober's (1984) example of short and tall organisms in a group-structured population, which we used to introduce cross-level by-products.) Situations in which organismic and cell fitness have not been decoupled, and where organisms thus lack true individuality, are precisely those in which there is a cross-level by-product running in the cell→organism direction. For as we saw in Chapter 3, particle→collective by-products, in an MLS2 framework, occur when the fitness of a collective is directly determined by the average fitness of the particles it contains.[9] And this is precisely the case in the example above. Given that C cells divide faster than D cells, the average cell fitness of an organism that starts as a CCC propagule is greater than the average cell fitness of one that starts as a DDD propagule; and it is this that explains why the former has higher organismic fitness. So the character-fitness covariance at the organism level will disappear when we control for average cell fitness.

Thirdly, this example has implications for Michod's use of the Price equation to model the evolution of multicellularity. As we know, where cross-level by-products are in play, the Price equation will detect selection at the higher level even when, intuitively, all the causality is at the

---

[9] Recall that this is a sufficient condition for an MLS2 particle→collective by-product, not a necessary condition.

lower level. This is precisely the case in the example above—all the selection is at the cellular level, so the character-fitness covariance at the organism level is spurious, or non-causal. Despite this, Michod makes extensive use of the Price equation in relation to this model, as a way of representing the effects of organismic and cellular selection on the overall change in frequency of the C allele/cell-type (Michod 1999; Michod and Roze 1999). Interestingly, however, in some of their papers Michod and Roze employ a different partitioning technique, which avoids the problem with the Price equation (Michod and Roze 2000; Roze and Michod 2001). This latter technique yields a *temporal* partition of the total change, into a component that happens during development, and a component that happens later, during reproduction (Figure 8.3). Michod and Roze then attribute these two components to selection at the cellular and organismic levels, respectively. This implies that if organismic fitness depends only on adult size, rather than adult functionality, all the selection is at the cellular level, for reproduction will not change the proportion of C cells in the population. Intuitively this is the correct result, but it is not compatible with taking organismic selection to be defined by the character-fitness covariance at the organism level, à la Price.

My purpose in pointing this out is not to criticize Michod and Roze, whose concept of fitness decoupling is surely invaluable for understanding evolutionary transitions. Rather, it is to emphasize the striking fact that the theoretical shortcomings of the Price approach, first pointed out by Sober (1984), Nunney (1985a), and Heisler and Damuth (1987) in relation to group selection, should reappear in relation to the evolution of multicellularity. It is particularly striking that the problem cases for the Price approach, where cross-level by-products are in play, represent transitional stages en route to the evolution of new hierarchical levels. The original critiques of the Price approach made no mention of this point, since they operated with a synchronic rather than a diachronic formulation of the levels-of-selection question.

8.5 CONCLUDING REMARKS

Recall our original question: which type of multi-level selection is relevant to evolutionary transitions? We have seen that MLS1 is relevant to the early stages of a transition, when the particles are just beginning to engage in cooperative interactions, while MLS2 is relevant later on, when a well-defined process of collective reproduction is in place,

Levels of Selection and the Major Evolutionary Transitions   237

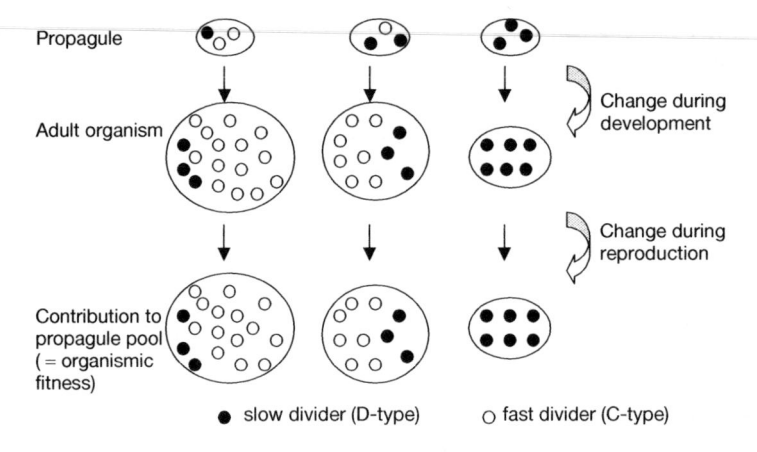

Figure 8.3. Differences in organismic fitness due solely to differences in cell division rates (based on Roze and Michod 2001 figure 7)

Note that for each organism, the second and third row circles are equal in size, that is, an organism's fitness is proportional to its adult size. This implies that all the evolutionary change happens during the development phase, and none during the reproduction phase. Roze and Michod (2001) take this to mean that there is no selection at the organismic level. However, on the Price approach there clearly is a component of organism-level selection—organismic fitness covaries positively with frequency of C-cells in the propagule. On the Price approach, 'no organism level selection' would require all the third row circles to be equal in size *to each other*, rather than equal in size to the second row circles.

so collective fitness in the MLS2 sense can apply. If the moral of Michod's model is generalizable to all evolutionary transitions, something more precise can be said about how this shift from MLS1 to MLS2 occurs.

Michod's model suggests that there will be a transitional phase during which collective fitness in the MLS2 sense can apply, that is, collectives do produce offspring collectives, but where a collective's fitness is directly dependent on the average fitness of its constituent particles, as in Figure 8.3. This represents a sort of grey area between MLS1 and MLS2: collective fitness is not *defined* as average particle fitness, as in MLS1, but it is *proportional* to average particle fitness; so the entirety of the collective-level character-fitness covariance is due to a cross-level by-product. In Michod's terms, this means that the emerging collective lacks 'individuality', and has no collective-level functions of its own. As the transition proceeds, collective fitness is gradually decoupled from average particle fitness, and starts to depend on the functionality

Table 8.1. Relation between collective fitness and particle fitness during an evolutionary transition

Stage 1	collective fitness defined as average particle fitness (cooperation spreads among particles)
Stage 2	collective fitness is not *defined* as average particle fitness but is *proportional* to average particle fitness (collectives start to emerge as entities in their own right with life cycles of their own)
Stage 3	collective fitness neither defined as nor proportional to average particle fitness (collectives have fully emerged; fitnesses are decoupled)

of the collective itself. MLS2 then occurs autonomously of MLS1, and the collectives can evolve adaptations of their own. Therefore, the relation between particle fitness and collective fitness *itself* undergoes a change, during a major evolutionary transition. This three-stage process is summarized in Table 8.1.

In a recent paper, Michod (2005) offers an interesting account of what makes an evolutionary transition complete. He argues that when the collective has finally emerged as a Darwinian individual, its average particle fitness will be zero. The basis for this idea is that the fitness of any biological unit equals the product of its viability and its fecundity; since evolutionary transitions typically involve reproductive division of labour, each of the particles in a fully formed collective either has zero viability, for example, a germ cell in a multicelled organism, or zero fecundity, for example, a somatic cell. So once the transition is complete, a collective's fitness in the MLS1 sense is zero, but its fitness in the MLS2 sense may be quite high, Michod argues.

This idea is valuable in that it emphasizes the importance of reproductive specialization in driving evolutionary transitions. However, in many cases it is natural to think of the particles as fitness-bearing units, even once the transition is effectively complete. For example, cancer in modern multicelled organisms is often regarded as a form of selfishness at the cellular level, and apoptosis (programmed cell death) as a form of altruism; these conceptualizations only make sense if individual cells are fitness-bearing units. Similarly, the insects in a eusocial insect colony are often treated as fitness-bearing units, for example, for the purposes of analysing intra-colonial conflicts of interest. So Michod's idea that an evolutionary transition is not complete until the particles have ceased altogether to be bearers of fitness seems overly restrictive.

Levels of Selection and the Major Evolutionary Transitions   239

If the three-stage schema above is a broadly correct picture of how evolutionary transitions occur, this has interesting implications for a number of issues discussed in previous chapters. Consider for example the distinction between MLS1 and MLS2 itself. When Damuth and Heisler (1988) introduced this distinction, they took themselves to be distinguishing between two different types of evolutionary process, both deserving of the label 'multi-level selection', which previous theorists had conflated. Their paradigm of MLS1 was traditional group selection and of MLS2 was species selection; they emphasized, correctly, that these selection processes are structurally different in key respects. But in the context of evolutionary transitions, the MLS1/MLS2 distinction assumes additional significance. Rather than simply describing selection processes of different sorts, which should be kept separate in the interests of conceptual clarity, MLS1 and MLS2 represent different *temporal* stages of an evolutionary transition. Damuth and Heisler did not appreciate this point, since they formulated the levels-of-selection question synchronically, rather than as a question about how the biological hierarchy first evolved.

In previous chapters, we discussed various examples where theorists disagree about how the level(s) of selection should be identified, or about whether a single-level or multi-level description of the selection process is preferable. Trait-group models, kin selection, and evolutionary game theory are cases in point. The disagreements surrounding these modes of selection have usually arisen in a synchronic context, where evolutionary transitions are not at issue. But in a diachronic context, the disagreements look different. For the modes of selection in question all involve the evolution of social behaviour, so are likely to have been important in evolutionary transitions. This makes the disagreements surrounding their status intelligible. For borderline cases of part–whole structure are inevitable in an evolutionary transition, given the gradualness of the Darwinian process. There is bound to be a period of indeterminacy during which it is unclear whether genuine collectives exist or not; how to classify selection processes that occur during this period will be similarly indeterminate. It is no accident that many of the most persistent disagreements over the level(s) of selection concern processes likely to have been important during these transitional phases. That such disagreements can persist, even when the basic empirical facts are not in dispute, is readily understood from a diachronic perspective.

I suspect that there is a general moral here. The conceptual issues that form the core of the traditional levels-of-selection debate, in both the biological and philosophical literatures, are subtly transformed when we move from a synchronic to a diachronic formulation of the levels question, as we must do if we are to understand the evolutionary transitions in Darwinian terms. But the transformation is not so drastic that the traditional discussions lose all their relevance.

To conclude, the analysis of evolutionary transitions offered in this chapter is very far from complete. I have not attempted a comprehensive survey of the field, but have focused on thematic and conceptual issues, particularly as they relate to the levels of selection. The study of evolutionary transitions is still in its infancy, with much empirical work remaining to be done, so it is difficult to say whether the foregoing analysis will prove satisfactory in all respects. But whatever future developments in the field look like, it is likely that multi-level selection will remain crucial for theorizing about evolutionary transitions. Given that multi-level selection theory, in its current form, is the product of many decades of work on the levels-of-selection problem in biology, it follows that this work has left an important intellectual legacy. C.H. Waddington famously dismissed the levels-of-selection debate of the 1960s as 'a rather foolish controversy', an opinion in which he was not alone.[10] The centrality of the levels-of-selection issue in the contemporary literature on evolutionary transitions shows that Waddington's opinion was seriously mistaken.

[10] Waddington's remark is quoted in Maynard Smith (1976), p. 277.

Bibliography

Alexander, R. D., and Borgia, G. (1978) 'Group Selection, Altruism, and the Levels of Organization of Life', *Annual Review of Ecology and Systematics* 9, 449–74.
Allee, W. C., Emerson, A., Park, O., Park, T., and Schmidt, K. (1949) *Principles of Animal Ecology*, Philadelphia: W. B. Saunders.
Arnold, A. J., and Fristrup, K. (1982) 'The Theory of Evolution by Natural Selection: A Hierarchical Expansion', *Paleobiology* 8, 113–29, reprinted in R. N. Brandon and R. M. Burian (eds.) (1984) *Genes, Organisms, Populations: Controversies over the Units of Selection*, Cambridge MA: MIT Press, 292–319.
Aspi, J., Jakalaniemi, A., Tuomi J., and Siikamaki, P. (2003) 'Multi-level Phenotypic Selection on Morphological Characters in a Metapopulation of *Silene Tatarica*', *Evolution* 57, 3, 509–17.
Avital, E., and Jablonka, E. (2000) *Animal Traditions: Behavioural Inheritance in Evolution*, Cambridge: Cambridge University Press.
Axelrod, R., and Hamilton, W. D. (1981) 'The Evolution of Cooperation', *Science* 211, 1390–6.
Barrett, M., and Godfrey-Smith, P. (2002) 'Group Selection, Pluralism, and the Evolution of Altruism', *Philosophy and Phenomenological Research* LXV, 3, 685–91.
Beatty, J. (1984) 'Chance and Natural Selection', *Philosophy of Science* 51, 183–211.
_____ (1992) 'Random Drift', in E. Keller and E. A. Lloyd (eds.) *Keywords in Evolutionary Biology*, Cambridge MA: Harvard University Press, 273–81.
Beckermann, A., Flohr, H., and Kim, J. (eds.) (1992) *Emergence or Reduction? Essays on the Prospects of Nonreductive Physicalism*, Berlin: de Gruyter.
Boorman, S. A., and Levitt, P. R. (1973) 'Group Selection at the Boundary of a Stable Population', *Theoretical Population Biology* 4, 85–128.
Boyd, L. H., and Iversen, G. R. (1979) *Contextual Analysis: Concepts and Statistical Techniques*, Belmont CA: Wadsworth.
Boyd, R., and Richerson, P. J. (1985) *Culture and the Evolutionary Process*, Chicago: University of Chicago Press.
_____ (2005) *The Origin and Evolution of Cultures*, Oxford: Oxford University Press.
Brandon, R. N. (1982) 'Levels of Selection', *Proceedings of the Philosophy of Science Association* vol. 1, East Lansing, MI: Philosophy of Science Association, 315–23.

Brandon, R. N. (1988) 'The Levels of Selection: A Hierarchy of Interactors', in H. C. Plotkin (ed.) *The Role of Behaviour in Evolution*, Cambridge MA: MIT Press, 51–71.

―― (1990) *Adaptation and Environment*, Princeton: Princeton University Press.

―― and Carson, S. (1996) 'The Indeterministic Character of Evolutionary Theory', *Philosophy of Science* 63, 315–37.

―― Antonovics, J., Burian, R. M., Carson, S., Coper, G., Davies, P. S., Hovarth, C., Mishler, B. D., Richardson, R. C., Smith, S., and Thrall, P. H. (1994) 'Sober on Brandon on Screening Off and the Levels of Selection', *Philosophy of Science* 61, 475–86.

Burt, A., and Trivers, R. (2006) *Genes in Conflict*, Cambridge MA: Harvard University Press.

Buss, L. W. (1987) *The Evolution of Individuality*, Princeton: Princeton University Press.

Clayton, P., and Davies, P. (eds.) (2006) *The Re-Emergence of Emergence*, Oxford: Oxford University Press.

Colwell, R. K. (1981) 'Group Selection is Implicated in the Evolution of Female-Biased Sex Ratios', *Nature* 290, 401–4.

Cosmides, L. M., and Tooby, J. (1981) 'Cytoplasmic Inheritance and Intragenomic Conflict', *Journal of Theoretical Biology* 89, 83–129.

Cronin, H. (2005) 'Adaptation: A Critique of Some Current Evolutionary Thought', *Quarterly Review of Biology* 80, 1, 19–27.

Crow, J. F. (1986) *Basic Concepts in Population, Quantitative and Evolutionary Genetics*, San Francisco: W. H. Freeman.

Damuth, J. (1985) 'Selection among "Species": A Formulation in terms of Natural Functional Units', *Evolution* 39, 5, 1132–46.

―― and Heisler, I. L. (1988) 'Alternative Formulations of Multi-level Selection', *Biology and Philosophy* 3, 407–30.

Darwin, C. (1859) *On the Origin of Species by Means of Natural Selection*, London: John Murray.

―― (1871) *The Descent of Man and Selection in Relation to Sex*, London: John Murray.

Dawkins, R. (1976) *The Selfish Gene*, Oxford: Oxford University Press.

―― (1982) *The Extended Phenotype*, Oxford: Oxford University Press.

―― (1984) 'Replicators and Vehicles', in R. N. Brandon and R. Burian (eds.) *Genes, Organisms, Populations: Controversies over the Units of Selection*, Cambridge MA, MIT Press, 161–80.

―― (2004) 'Extended Phenotype—But Not Too Extended. A Reply to Laland, Turner and Jablonka', *Biology and Philosophy* 19, 3, 377–96.

Dennett, D. C. (1995) *Darwin's Dangerous Idea*, London: Penguin.

―― (2002) 'Commentary on Sober and Wilson's *Unto Others*', *Philosophy and Phenomenological Research* LXV, 3, 692–7.

Dobzhansky, T. (1951) *Genetics and the Origin of Species*, 3rd edn., New York: Columbia University Press.

Doolittle, W. F., and Sapienza, C. (1980) 'Selfish Genes, the Phenotype Paradigm and Genome Evolution', *Nature* 284, 601–3.

Dugatkin, L. A., and Reeve, H. K. (1994) 'Behavioural Ecology and Levels of Selection: Dissolving the Group Selection Controversy', *Advances in the Study of Behaviour* 23, 101–33.

Edwards, A.W. F. (1994) 'The Fundamental Theorem of Natural Selection', *Biological Reviews* 69, 443–74.

Eigen, M., and Schuster, P. (1977) 'The Hypercycle, a Principle of Natural Self-Organization. Part A: Emergence of the Hypercycle', *Naturwissenschaften* 64, 541–65.

____ (1979) *The Hypercycle, a Principle of Natural Self-Organization*, Berlin: Springer-Verlag.

Eldredge, N. (1985) *Unfinished Synthesis: Biological Hierarchies and Modern Evolutionary Thought*, New York: Oxford University Press.

____ (1989) *Macroevolutionary Dynamics: Species, Niches and Adaptive Peaks*, New York: McGraw-Hill.

____ (2003) 'The Sloshing Bucket: How the Physical Realm Controls Evolution' in J. Crutchfield and P. Schuster (eds.) *Evolutionary Dynamics*, Oxford: Oxford University Press, 3–32.

____ and Gould, S. J. (1972) 'Punctuated Equilibria: an Alternative to Phyletic Gradualism', in T. J. M. Schopf (ed.) *Models in Paleobiology*, San Francisco: Freeman, 82–215.

Elster, J. (1982) 'The Case for Methodological Individualism', *Theory and Society* 11, 453–82.

Emerson, A. E. (1939) 'Social Coordination and the Superorganism', *American Naturalist* 21, 182–209.

Endler, J. A. (1986) *Natural Selection in the Wild*, Princeton: Princeton University Press.

Ereshefsky, M. (ed.) (1992) *The Units of Evolution: Essays on the Nature of Species,* Cambridge MA: MIT Press.

Ewens, W. (1989) 'An Interpretation and Proof of the Fundamental Theorem of Natural Selection', *Theoretical Population Biology* 36, 167–80.

Fagerstrom, T. (1992) 'The Meristem Cycle as a Basis for Defining Fitness in Clonal Plants', *Oikos* 63, 449–53.

Fagerstrom, T., Briscoe, D., and Sunnucks, P. (1998) 'Evolution of Mitotic Cell Lineages', *Trends in Ecology and Evolution* 13, 3, 117–20.

Falconer, D. S. (1989) *Introduction to Quantitative Genetics*, 3rd edn., London: Longman.

Falk, R., and Sarkar, S. (1992) 'Harmony from Discord', *Biology and Philosophy* 7, 463–72.

Fisher, R. A. (1930) *The Genetical Theory of Natural Selection*, Oxford: Clarendon Press.
Fletcher, J. A., and Zwick, M. (2004) 'Strong Altruism can Evolve in Randomly Formed Groups', *Journal of Theoretical Biology* 228, 303–13.
Fontana, W., and Buss, L. W. (1994) 'What Would be Conserved if the Tape Were Played Twice?', *Proceedings of the National Academy of Sciences* 91, 757–61.
Frank, S. A. (1995a) 'George Price's Contribution to Evolutionary Genetics', *Journal of Theoretical Biology* 175, 373–88.
—— (1995b) 'Mutual Policing and Repression of Competition in the Evolution of Co-operative Groups', *Nature* 377, 520–2.
—— (1997) 'The Price Equation, Fisher's Fundamental Theorem, Kin Selection and Causal Analysis', *Evolution* 51, 1712–29.
—— (1998) *Foundations of Social Evolution*, Princeton: Princeton University Press.
Getty, T. (1999) 'What do Experimental Studies Tell Us about Group Selection in Nature?', *American Naturalist* 154, 596–8.
Ghiselin, M. T. (1974a) 'A Radical Solution to the Species Problem', *Systematic Zoology* 23, 536–44.
—— (1974b) *The Economy of Nature and the Evolution of Sex*, Berkeley: University of California Press.
Gilinsky, N. (1986) 'Species Selection as a Causal Process', *Evolutionary Biology* 20, 249–73.
Godfrey-Smith, P. (1992) 'Additivity and the Units of Selection', in D. Hull, M. Forbes, and K. Okruhlik (eds.) *PSA 1992*, Vol 1, East Lansing: Philosophy of Science Association, 315–28.
—— (2000a) 'The Replicator in Retrospect', *Biology and Philosophy* 15, 403–23.
—— (2000b) 'Information, Arbitrariness and Selection: Comments on Maynard Smith', *Philosophy of Science* 67, 2, 202–7.
—— (forthcoming) 'Varieties of Population Structure and the Levels of Selection'.
—— and Kerr, B. (2002) 'Group Fitness and Multi-level Selection: Replies to Commentaries', *Biology and Philosophy* 17, 4, 539–49.
Goodnight, C. J., Schwartz, J. M., and Stevens, L. (1992) 'Contextual Analysis of Models of Group Selection, Soft Selection, Hard Selection, and the Evolution of Altruism', *American Naturalist* 140, 743–61.
Gould, S. J. (1982) 'The Meaning of Punctuated Equilibrium and its Role in Validating a Hierarchical Approach to Macroevolution', in R. Milkman (eds.) *Perspectives on Evolution*, Sunderland MA: Sinauer, 83–104.
—— (1984) 'Caring Groups, Selfish Genes', in E. Sober (ed.) *Conceptual Issues in Evolutionary Biology*, Cambridge MA: MIT Press, 119–24.

_____ (2001) 'The Evolutionary Definition of Selective Agency, Validation of the Theory of Hierarchical Selection, and Fallacy of the Selfish Gene', in R. S. Singh, K. Krimbas, D. Paul, and J. Beatty (eds.) *Thinking about Evolution: Historical, Philosophical and Political Perspectives* vol. 2, Cambridge: Cambridge University Press, 207–29.

_____ (2002) *The Structure of Evolutionary Theory*, Cambridge MA: Harvard University Press.

_____ and Eldredge, N. (1977) 'Punctuated Equilibrium: the Tempo and Mode of Evolution Reconsidered', *Paleobiology* 3, 115–51.

_____ and Lloyd, E. A. (1999) 'Individuality and Adaptation across Levels of Selection', *Proceedings of the National Academy of Sciences* 96, 21, 11904–9.

Grafen, A. (1984) 'Natural Selection, Kin Selection and Group Selection', in J. R. Krebs and N. B. Davies (eds.) *Behavioural Ecology: an Evolutionary Approach*, Oxford: Blackwell, 62–84.

_____ (1985) 'A Geometric View of Relatedness', *Oxford Surveys in Evolutionary Biology* 2: 28–89.

_____ (2000) 'Developments of the Price Equation and Natural Selection under Uncertainty', *Proceedings of the Royal Society of London B* 267, 1223–7.

_____ (2003) 'Fisher the Evolutionary Biologist', *Journal of the Royal Statistical Society*, Series D, 52, 319–29.

Grantham, T. A. (1995) 'Hierarchical Approaches to Macroevolution: Recent Work on Species Selection and the Effect Hypothesis', *Annual Review of Ecology and Systematics* 26, 301–21.

Griesemer, J. (2000) 'The Units of Evolutionary Transition', *Selection* 1, 67–80.

Haber, M., and Hamilton, A. (forthcoming) 'Coherence, Consistency and Cohesion: Clade Selection in Okasha and Beyond', *Philosophy of Science (Proceedings)*.

Haig, D., and Grafen, A. (1991) 'Genetic Scrambling as a Defence against Meiotic Drive', *Journal of Theoretical Biology* 153, 531–58.

Haldane, J. B. S. (1932) *The Causes of Evolution*, London: Longman.

_____ (1964) 'A Defense of Beanbag Genetics', *Perspectives in Biology and Medicine* 19, 343–59.

Hamilton, W. D. (1963) 'The Evolution of Altruistic Behaviour', *The American Naturalist* 97, 354–6, reprinted in Hamilton (1996), 6–8.

_____ (1964) 'The Genetical Evolution of Social Behaviour I', *Journal of Theoretical Biology* 7, 1–16, reprinted in Hamilton (1996), 31–46.

_____ (1967) 'Extraordinary Sex Ratios', *Science* 156, 477–88, reprinted in Hamilton (1996), 143–69.

_____ (1975) 'Innate Social Aptitudes in Man: an Approach from Evolutionary Genetics', in R. Fox (ed.) *Biosocial Anthropology*, New York: Wiley, 133–55, reprinted in Hamilton (1996), 315–52.

_____ (1996) *Narrow Roads of Gene Land. Volume 1: Evolution of Social Behaviour*. New York: W. H. Freeman.

Hammerstein, P. (2003) 'Why is Reciprocity so Rare in Social Animals? A Protestant Appeal', in Hammerstein (ed.) (2003), 83–94.

―― (ed.) (2003) *Genetic and Cultural Evolution of Cooperation*, Cambridge MA: MIT Press.

Hansen, T. A. (1983) 'Modes of Larval Development and Rates of Speciation in early Tertiary Neogastropods', *Science* 220, 501–2.

Heinrich, J., Bowles, S., Boyd, R., Hopfensitz, A., Richerson, P. J., Sigmund, K., Smith, E. A., Weissing, F. J., and Young, H. P. (2003) 'Group Report: The Cultural and Genetic Evolution of Human Cooperation', in Hammerstein (ed.) (2003), 445–68.

Heisler, I. L. and Damuth, J. (1987) 'A Method for Analyzing Selection in Hierarchically Structured Populations', *American Naturalist* 130, 582–602.

Hennig, W. (1966) *Phylogenetic Systematics*, Urbana: University of Illinois Press.

Heywood, J. S. (2005) 'An Exact Form of the Breeder's Equation for the Evolution of a Quantitative Trait under Natural Selection', *Evolution* 59, 11, 2287–98.

Hodge, M. J. S. (1992) 'Biology and Philosophy (including Ideology): A Study of Fisher and Wright', in S. Sarkar (ed.) *The Founders of Evolutionary Genetics: A Centenary Reappraisal*, Dordrecht: Kluwer, 231–93.

Hoekstra, R. F. (1990) 'Evolution of Uniparental Inheritance of Cytoplasmic DNA', in J. Maynard Smith and G. Vida (eds.) *Organizational Constraints on the Dynamics of Evolution*, New York: Manchester University Press, 269–78.

―― (2003) 'Power in the Genome: Who Suppresses the Outlaw?', in P. Hammerstein (ed.) *Genetic and Cultural Evolution of Cooperation*, Cambridge MA: MIT Press, 257–71.

Hoffman, A., and Hecht, M. K. (1986) 'Species Selection as a Causal Process: a Reply', *Evolutionary Biology* 20, 275–81.

Hull, D. L. (1978) 'A Matter of Individuality', *Philosophy of Science* 45, 335–60.

―― (1981) 'Units of Evolution: a Metaphysical Essay', in U. J. Jensen and R. Harré (eds.) *The Philosophy of Evolution*, 23–44. Brighton: Harvester Press.

Hurst, G. D. and Werren, J. (2001) 'The Role of Selfish Genetic Elements in Eukaryotic Evolution', *Nature Reviews* 2, 597–602.

Hurst, L. D. (1993) 'The Incidences, Mechanisms and Evolution of Cytoplasmic Sex Ratio Distorters in Animals', *Biological Reviews* 68, 121–93.

―― (1996) 'Adaptation and Selection of Genomic Parasites', in M. R. Rose and G. V. Lauder (eds.) *Adaptation*, San Diego: Academic Press, 407–49.

―― and Hamilton, W. D. (1992) 'Cytoplasmic Fusion and the Nature of the Sexes', *Proceedings of the Royal Society of London B* 247, 189–94.

―― Atlan, A., and Bengtsson, B. (1996) 'Genetic Conflicts', *Quarterly Review of Biology* 71, 317–64.

Irzik, G., and Meyer, E. (1987) 'Causal Modeling: New Directions for Statistical Explanation', *Philosophy of Science* 54, 495–514.

Jablonka, E. (1994) 'Inheritance Systems and the Evolution of New Levels of Individuality', *Journal of Theoretical Biology* 170, 301–9.
____ (2001) 'The Systems of Inheritance', in S. Oyama, P. Griffiths, and R. D. Gray (eds.) *Cycles of Contingency*, Cambridge MA: MIT Press, 99–116.
____ and Lamb, M. (1995) *Epigenetic Inheritance and Evolution: the Lamarckian Dimension*, Oxford: Oxford University Press.
Jablonski, D. (1982) 'Evolutionary Rates and Modes in Late Cretaceous Gastropods: Role of Larval Ecology', *Proceedings of the 3rd North American Paleontology Convention*, vol. 1, 257–62.
____ (1986) 'Larval Ecology and Macroevolution in Marine Invertebrates', *Bulletin of Marine Science* 39, 565–87.
____ (1987) 'Heritability at the Species Level: Analysis of Geographic Ranges of Cretaceous Mollusks', *Science* 238, 360–3.
Janzen, D. H. (1977) 'What are Dandelions and Aphids?', *American Naturalist* 111, 586–9.
Keller, L. (ed.) (1999) *Levels of Selection in Evolution*, Princeton: Princeton University Press.
Kerr, B., and Godfrey-Smith, P. (2002) 'Individualist and Multi-level Perspectives on Selection in Structured Populations', *Biology and Philosophy* 17: 477–517.
____ and Feldman, M. (2004) 'What is Altruism?', *Trends in Ecology and Evolution*, 19, 135–40.
Kim, J. (1998) *Mind in a Physical World*, Cambridge MA: MIT Press.
Kitcher, P. (2004) 'Interview with Philip Kitcher', *Human Nature Review* 4, 87–92,<http://human-nature.com/nibbs/04/kitcher.html>.
____ Sterelny, K., and Waters, C. K. (1990) 'The Illusory Riches of Sober's Monism', *Journal of Philosophy* 87, 158–61.
Krause, J., and Ruxton, G. D. (2002) *Living in Groups*, Oxford: Oxford University Press.
Lack, D. (1966) *Population Studies of Birds*, Oxford: Clarendon Press.
Lande, R., and Arnold, S. J. (1983) 'The Measurement of Selection on Correlated Characters', *Evolution*, 37, 1210–26.
Leigh, E. G. Jr. (1977) 'How does Natural Selection Reconcile Individual Advantage with the Good of the Group?', *Proceedings of the National Academy of the Sciences USA* 74, 4524–46.
Levin, B. R., and Kilmer, W. C. (1974) 'Interdemic Selection and the Evolution of Altruism: a Computer Simulation Study', *Evolution* 28, 527–45.
Levins, R. (1970) 'Extinction' in M. Gerstenbaher (ed.) (1970) *Some Mathematical Questions in Biology*, Providence, RI: American Mathematical Society, 77–107.
Lewis, D. (1973) 'Causation', *Journal of Philosophy* 70, 556–67.
Lewontin, R. C. (1970) 'The Units of Selection', *Annual Review of Ecology and Systematics* 1, 1–18.

Lewontin, R. C. (1991) 'Review of E. Lloyd, The Structure and Confirmation of Evolutionary Theory', *Biology and Philosophy* 6, 461–6.
―― (1998) 'Survival of the Nicest? Review of *Unto Others* by E. Sober and D. S. Wilson', *New York Review of Books* 22, 59–63.
―― (2000) *It Ain't Necessarily So: The Dream of the Human Genome and Other Illusions*, New York: New York Review Books.
Lieberman, B. S. (1995) 'Phylogenetic Trends and Speciation: Analyzing Macroevolutionary Processes and Levels of Selection', in D. H. Erwin and R. L. Anstey (eds.) *Speciation in the Fossil Record*, New York: Columbia University Press, 316–39.
―― and Vrba, E. S. (1995) 'Hierarchy Theory, Selection and Sorting', *Bioscience* 45, 6, 394–9.
Lloyd, E. A. (1988) *The Structure and Confirmation of Evolutionary Theory*, New York: Greenwood Press.
―― and Gould, S. J. (1993) 'Species Selection on Variability', *Proceedings of the National Academy of Sciences* 90, 595–9.
Loewer, B. (2002) 'Comments on Jaegwon Kim's *Mind in a Physical World*', *Philosophy and Phenomenological Research* LXV, 3, 655–62.
Lorenz, K. (1963) *On Aggression*, San Diego: Harcourt Brace.
Lukes, S. (1968) 'Methodological Individualism Reconsidered', *The British Journal of Sociology* 19, 119–29.
Lyttle, T. W. (1991) 'Segregation Distorters', *Annual Review of Genetics* 25, 511–57.
McShea, D. W. (1996) 'Metazoan Complexity and Evolution: is there a Trend?', *Evolution* 50, 477–92.
―― (1998) 'Possible Large-Scale Trends in Organismal Evolution: Eight Live Hypotheses', *Annual Review of Ecology and Systematics* 29, 293–318.
―― (2001a) 'The "Minor Transitions" in Hierarchical Evolution and the Question of Directional Bias', *Journal of Evolutionary Biology* 14, 502–18.
―― (2001b) 'The Hierarchical Structure of Organisms: a Scale and Documentation of a Trend in the Maximum', *Paleobiology* 27, 405–23.
Margulis, L. (1981) *Symbiosis in Cell Evolution*, San Francisco: W. H. Freeman.
Matessi, C., and Jayakar, S. D. (1976), 'Conditions for the Evolution of Altruism under Darwinian Selection', *Theoretical Population Biology* 9, 360–87.
Matthen, M., and Ariew, A. (2002) 'Two Ways of Thinking About Fitness and Natural Selection', *Journal of Philosophy* 49, 55–83.
Maynard Smith, J. (1964) 'Group Selection and Kin Selection', *Nature* 201, 1145–7.
―― (1966) *The Theory of Evolution* 2nd edn., London: Penguin.
―― (1976) 'Group Selection', *Quarterly Review of Biology* 51, 277–83.
―― (1978) *The Evolution of Sex*, Cambridge: Cambridge University Press.
―― (1982) *Evolution and the Theory of Games*, Cambridge: Cambridge University Press.

___ (1983) 'Current Controversies in Evolutionary Biology', in M. Grene (ed.) *Dimensions of Darwinism*, Cambridge: Cambridge University Press, 273–86.

___ (1987a) 'How to Model Evolution', in J. Dupré (ed.) *The Latest on the Best: Essays on Evolution and Optimality*, Cambridge MA: MIT Press, 119–31.

___ (1987b) 'Reply to Sober', in J. Dupré (ed.) *The Latest on the Best: Essays on Evolution and Optimality*, Cambridge MA: MIT Press, 147–50.

___ (1998) 'The Origin of Altruism', *Nature* 393, 639–40.

___ (2000) 'The Concept of Information in Biology', *Philosophy of Science*, 67, 2, 177–94.

___ (2002) 'Commentary on Kerr and Godfrey-Smith', *Biology and Philosophy* 17, 4, 523–7.

___ and Price, G. R. (1973) 'The Logic of Animal Conflict', *Nature* 246, 15–18.

___ and Szathmáry, E. (1995) *The Major Transitions in Evolution*, Oxford: Oxford University Press.

Mayo, D. G., and Gilinsky, N. L. (1987) 'Models of Group Selection', *Philosophy of Science* 54, 515–38.

Mayr, E. (1959) 'Where Are We?', *Cold Spring Harbour Symposium on Quantitative Biology* 24, 1–14.

___ (1963) *Animal Species and Evolution*, Cambridge MA: Harvard University Press.

Michod, R. E. (1982) 'The Theory of Kin Selection', *Annual Review of Ecology and Systematics* 13, 23–55.

___ (1983) 'Population Biology of the First Replicators: On the Origin of the Genotype, Phenotype and Organism', *American Zoologist* 23, 5–14.

___ (1997) 'Cooperation and Conflict in the Evolution of Individuality I. Multilevel Selection of the Organism', *American Naturalist* 149, 607–45.

___ (1998) 'Evolution of Individuality', *Journal of Evolutionary Biology* 11, 225–7.

___ (1999) *Darwinian Dynamics: Evolutionary Transitions in Fitness and Individuality*, Princeton: Princeton University Press.

___ (2000) 'Some Aspects of Reproductive Mode and the Origin of Multicellularity', *Selection* 1, 97–109.

___ (2005) 'On the Transfer of Fitness from the Cell to the Organism', *Biology and Philosophy* (forthcoming).

___ and Nedelcu, A. (2003) 'On the Reorganization of Fitness During Evolutionary Transitions in Individuality', *Integrative and Comparative Biology* 43, 64–73.

___ and Roze, D. (1999) 'Cooperation and Conflict in the Evolution of Individuality III', in C. L. Nehaniv (ed.) *Mathematical and Computational Biology: Computational Morphogenesis, Hierarchical Complexity, and Digital*

Evolution, vol. 26, American Mathematical Society, Providence, Rhode Island, 47–92.
Mills, S. K., and Beatty, J. (1979) 'The Propensity Interpretation of Fitness', *Philosophy of Science* 46, 263–86.
Mishler, B. D., and Brandon, R. N. (1987) 'Individuality, Pluralism and the Phylogenetic Species Concept', *Biology and Philosophy* 2, 397–414.
Monro, K., and Poore, A. G. (2004) 'Selection in Modular Organisms: Is Intraclonal Variation in Macroalgae Evolutionarily Important?', *American Naturalist* 163, 4, 564–78.
Nelson, G. J., and Platnick, N. (1984) 'Systematics and Evolution', in M. Ho and P. T. Saunders (eds.) *Beyond Neo-Darwinism*, London: Academic Press, 143–58.
Northcott, R. (2005) 'Comparing Apples with Oranges', *Analysis* 65, 12–18.
Nunney, L. (1985a) 'Group Selection, Altruism, and Structured-Deme Models', *American Naturalist* 126, 212–30.
____ (1985b) 'Female-biased Sex Ratios: Individual or Group Selection?' *Evolution* 39, 349–61.
____ (1993) 'Review of George C. Williams, *Natural Selection: Domains, Levels, Challenges*', *Journal of Evolutionary Biology* 6, 5, 773.
____ (1998) 'Are We Selfish, Are We Nice, or Are We Nice Because We Are Selfish?', *Science* 281, 5383, 1619–21.
____ (1999), 'Lineage Selection: Natural Selection for Long Term Benefit', in L. Keller (ed.) *Levels of Selection in Evolution*, Princeton: Princeton University Press, 238–52.
____ (2002) 'Altruism, Benevolence and Culture', in L. D. Katz (ed.) *Evolutionary Origins of Morality*, Thorverton: Imprint Academic, 231–6.
Odling-Smee, F. J., Laland, K. N., and Feldman, M. W. (2003) *Niche Construction: The Neglected Process in Evolution*, Princeton: Princeton University Press.
Okasha, S. (2001), 'Why Won't the Group Selection Controversy Go Away?', *British Journal for the Philosophy of Science* 52, 1, 25–50.
____ (2002) 'Genetic Relatedness and the Evolution of Altruism', *Philosophy of Science* 69, 1, 138–49.
____ (2003a) 'Does the Concept of "Clade Selection" Make Sense?', *Philosophy of Science* 70, 739–51.
____ (2003b)'The Concept of Group Heritability', *Biology and Philosophy* 18, 3, 445–61.
____ (2004a)'The "Averaging Fallacy" and the Levels of Selection', *Biology and Philosophy* 19, 167–84.
____ (2004b), 'Multi-level Selection and the Partitioning of Covariance: A Comparison of Three Approaches', *Evolution* 58, 3, 486–94.
____ (2004c), 'Multi-level Selection, Covariance and Contextual Analysis', *British Journal for the Philosophy of Science* 55, 481–504.

____ (2005) 'Altruism, Group Selection and Correlated Interaction', *British Journal for the Philosophy of Science* 56, 703–27.

____ (2006) 'Multi-level Selection and the Major Transitions in Evolution', *Philosophy of Science (Proceedings)* (forthcoming).

Orgel, L. E., and Crick, F. H. C. (1980) 'Selfish DNA: The Ultimate Parasite', *Nature* 284, 604–7.

Oyama, S. (2000) *Evolution's Eye: a Systems View of the Biology/Culture Divide*, Durham NC: Duke University Press.

Pan, J. J., and Price, J. S. (2002) 'Fitness and Evolution in Clonal Plants: The Impact of Clonal Growth', *Evolutionary Ecology* 15, 583–600.

Pearson, K. (1903) 'Mathematical Contributions to the Theory of Evolution XI. On the Influence of Natural Selection on the Variability and Correlation of Organs', *Philosophical Transactions of the Royal Society of London*, A200, 1–66.

Pedersen, B., and Tuomi, J. (1995) 'Hierarchical Selection and Fitness in Modular and Clonal Organisms', *Oikos* 73, 167–80.

Pomiankowski, A. (1999) 'Intragenomic Conflict', in L. Keller (ed.) *Levels of Selection in Evolution*, 121–52.

Price, G. R. (1972) 'Extension of Covariance Selection Mathematics', *Annals of Human Genetics* 35, 485–90.

____ (1995) 'The Nature of Selection', *Journal of Theoretical Biology* 175, 389–96 (written circa 1971).

Price, H. (2005) 'Causal Perpsectivalism', in H. Price and R. Corry (eds.) *Causation, Physics and the Constitution of Reality: Russell's Republic Revisited*, Oxford: Oxford University Press (forthcoming).

Queller, D. C. (1992a) 'A General Model for Kin Selection', *Evolution* 46, 376–80.

____ (1992b) 'Quantitative Genetics, Inclusive Fitness, and Group Selection', *American Naturalist* 139, 540–58.

____ (1997) 'Cooperators since Life Began', *Quarterly Review of Biology* 72, 184–8.

____ (2000) 'Relatedness and the Fraternal Major Transitions', *Philosophical Transactions of the Royal Society* B 355, 1647–55.

____ and Strassman, J. (2003) 'Eusociality', *Science* 13, 22, 861–3.

Raup, D. M., Gould S. J., Schopf, T. J. M., and Simberloff, D. S. (1973) 'Stochastic Models of Phylogeny and the Evolution of Diversity', *Journal of Geology* 81, 525–42.

Reeve, H. K., and Keller, L. (1999) 'Levels of Selection: Burying the Units-of-Selection Debate and Unearthing the Crucial New Issues', in L. Keller (ed.) *Levels of Selection in Evolution*, 3–14.

Reichenbach, H. (1956) *The Direction of Time*, Berkeley: University of California Press.

Rice, S. (2004) *Evolutionary Theory*, Sunderland MA: Sinauer.

Ridley, M., and Grafen, A. (1981) 'Are Green Beard Genes Outlaws?', *Animal Behaviour* 29, 954–5.

Rosenberg, A. (1994) *Instrumental Biology, or the Disunity of Science*, Chicago: University of Chicago Press.

Roughgarden, J. (1979) *Theory of Population Genetics and Evolutionary Ecology: An Introduction*, New York: Macmillan.

Roze, D., and Michod, R. E. (2001), 'Mutation, Multilevel Selection and the Evolution of Propagule Size during the Origin of Multicellularity', *American Naturalist* 158, 638–54.

Salmon, W. (1971) *Statistical Explanation and Statistical Relevance*, Pittsburgh: University of Pittsburgh Press.

Salt, G. W. (1979) 'A Comment on the Use of the Term "Emergent Properties"', *American Naturalist* 113, 145–8.

Sarkar, S. (1994) 'The Additivity of Variance and the Selection of Alleles', in D. Hull, M. Forbes, and R. Burian (eds.) *PSA 1994*, Vol 1, East Lansing: Philosophy of Science Association, 3–12.

____ (1998) *Genetics and Reductionism*, Cambridge: Cambridge University Press.

Seger, J. (1981) 'Kinship and Covariance', *Journal of Theoretical Biology* 91, 191–213.

Segerstråle, U. (2000) *Defenders of the Truth: The Battle for Science in the Sociobiology Debate and Beyond*, Oxford: Oxford University Press.

Shapiro, J. A., and Dworkin, M. (1997) *Bacteria as Multicellular Organisms*, Oxford: Oxford University Press.

Skyrms, B. (1995) *Evolution of the Social Contract*, Cambridge: Cambridge University Press.

____ (2002) 'Critical Commentary on Unto Others', *Philosophy and Phenomenological Research* LXV, 3, 697–701.

Sober, E. (1984) *The Nature of Selection*, Chicago: Chicago University Press.

____ (1987) 'Comments on Maynard Smith's "How to Model Evolution"', in J. Dupré (ed.) *The Latest on the Best: Essays on Evolution and Optimality*, Cambridge MA: MIT Press, 133–45.

____ (1988) 'Apportioning Causal Responsibility', *Journal of Philosophy* 85, 303–18.

____ (1992) 'Screening Off and the Units of Selection', *Philosophy of Science* 59, 142–52.

____ and Lewontin, R. C. (1982) 'Artifact, Cause and Genic Selection', *Philosophy of Science* 49, 157–80, reprinted in R. N. Brandon and R. Burian (1984) (eds.) *Genes, Organisms, Populations*, Cambridge MA: MIT Press, 109–32.

____ and Wilson, D. S. (1994) 'A Critical Review of Philosophical Work on the Units of Selection Problem', *Philosophy of Science* 61, 534–55.

Bibliography

―― (1998) *Unto Others: The Evolution and Psychology of Unselfish Behavior*, Cambridge MA: Harvard University Press.

―― (2002a) 'Perspectives and Parameterizations: Commentary on Benjamin Kerr and Peter Godfrey-Smith's "Individualist and Multi-level Perspectives on Selection in Structured Populations"', *Biology and Philosophy* 17, 4, 529–37.

―― (2002b) 'Morality and *Unto Others*', in L. D. Katz (ed.) *Evolutionary Origins of Morality*, Thorverton: Imprint Academic, 257–68.

―― (2002c) 'Reply to Commentaries', *Philosophy and Phenomenological Research* LXV, 3, 711–27.

Stanley, S. M. (1975) 'A Theory of Evolution above the Species Level', *Proceedings of the National Academy of Sciences* 72, 646–50.

―― (1979) *Macroevolution: Pattern and Process*, San Francisco: Freeman.

Steen, W. J. van der, and Berg, H. A. van den (1999) 'Dissolving Disputes over Genic Selectionism', *Journal of Evolutionary Biology* 12, 177–83.

Sterelny, K. (1996a) 'Explanatory Pluralism in Evolutionary Biology', *Biology and Philosophy* 11, 193–214.

―― (1996b) 'The Return of the Group', *Philosophy of Science* 63, 562–84.

―― (1999) 'Bacteria at the High Table', *Biology and Philosophy* 14, 459–70.

―― (2000) 'The "Genetic Program" Program: A Commentary on Maynard Smith on Information in Biology', *Philosophy of Science* 67, 2, 195–201.

―― (2006) 'Are Ecosystems Real?', *Philosophy of Science* (forthcoming).

―― and Griffiths, P. (1999) *Sex and Death: an Introduction to the Philosophy of Biology*, Chicago: Chicago University Press.

―― and Kitcher, P. (1988) 'The Return of the Gene', *Journal of Philosophy* 85, 339–60.

Stidd, B. M., and Wade, D. L. (1995) 'Is Species Selection Dependent upon Emergent Characters?', *Biology and Philosophy* 10, 55–76.

Szathmáry, E. and Demeter, L. (1987) 'Group Selection of Early Replicators and the Origin of Life', *Journal of Theoretical Biology* 128, 463–86.

―― and Maynard Smith, J. (1997), 'From Replicators to Reproducers: The First Major Transitions Leading to Life', *Journal of Theoretical Biology* 187, 555–71.

―― and Wolpert, L. (2003) 'The Transition from Single Cells to Multicellularity', in P. Hammerstein (ed.) *Genetic and Cultural Evolution of Cooperation*, Cambridge MA: MIT Press, 271–90.

Trivers, R. L. (1971) 'The Evolution of Reciprocal Altruism', *Quarterly Review of Biology* 46, 35–57.

―― (1985) *Social Evolution*, Menlo Park, CA: Benjamin Cummings.

Tsuji, K. (1995) 'Reproductive Conflicts in the Ant *Pristomyrmex Pungens*: Contextual Analysis and Partitioning of Covariance', *American Naturalist* 154, 599–613.

Tuomi, J., and Vuorisalo, T. (1989) 'Hierarchical Selection in Modular Organisms', *Trends in Ecology and Evolution* 4, 7, 209–13.

Uyenoyama, M. and Feldman, M. W. (1980) 'Theories of Kin and Group Selection: a Population Genetics Perspective', *Theoretical Population Biology* 17, 380–414.

——— (1992) 'Altruism: Some Theoretical Ambiguities' in E. F. Keller and E. A. Lloyd (eds.) *Keywords in Evolutionary Biology*, Harvard: Harvard University Press, 151–7.

Valen, L. van (1988) 'Species, Sets, and the Derivative Nature of Philosophy', *Biology and Philosophy* 3, 49–66.

Vermeij, G. J. (1996) 'Adaptations of Clades: Resistance and Response', in M. R. Rose and G. Lauder (eds.) *Adaptation*, San Diego: Academic Press, 363–80.

Vrba, E. (1984) 'What is Species Selection?', *Systematic Zoology* 33, 318–28.

——— (1989) 'Levels of Selection and Sorting with Special Reference to the Species Level', *Oxford Surveys in Evolutionary Biology* 6, 111–68.

——— and Gould, S. J. (1986) 'The Hierarchical Expansion of Sorting and Selection', *Paleobiology* 12, 217–28.

Vries, H. de (1905) *Species and Varieties: Their Origin by Mutation*, Chicago: Open Court.

Wade, M. (1978) 'A Critical Review of the Models of Group Selection', *Quarterly Review of Biology* 53, 101–14.

——— (1985) 'Soft Selection, Hard Selection, Kin Selection, and Group Selection', *American Naturalist* 125, 61–73.

——— and Kalisz, S. (1990) 'The Causes of Natural Selection', *Evolution* 44, 1947–55.

——— Goodnight, C. J., and Stevens, L. (1999) 'Design and Interpretation of Experimental Studies of Interdemic Selection: a Reply to Getty', *American Naturalist* 154, 599–603.

Walton, D. (1991) 'The Units of Selection and the Bases of Selection', *Philosophy of Science* 58, 417–35.

Weismann, A. (1903) *The Evolution Theory*, London: Edward Arnold.

West-Eberhard, M. J. (1975) 'The Evolution of Social Behaviour by Kin Selection', *Quarterly Review of Biology* 50, 1–33.

Williams, G. C. (1966) *Adaptation and Natural Selection*, Princeton: Princeton University Press.

——— (1985) 'A Defense of Reductionism in Evolutionary Biology', *Oxford Surveys in Evolutionary Biology* 2, 1–27.

——— (1992) *Natural Selection: Domains, Levels, Challenges*, Oxford: Oxford University Press.

Wilson, D. S. (1975) 'A Theory of Group Selection', *Proceedings of the National Academy of Sciences USA*, 72, 143–6.

―― (1977) 'Structured Demes and the Evolution of Group Advantageous Traits', *American Naturalist* 111, 157–85.

―― (1980) *The Natural Selection of Populations and Communities*, Menlo Park CA: Benjamin/Cummings.

―― (1983) 'The Group Selection Controversy: History and Current Status', *Annual Review of Ecology and Systematics* 14, 159–87.

―― (1987) 'Altruism in Mendelian Populations derived from Sibling Groups: the Haystack Model Revisited', *Evolution* 41, 1059–70.

―― (1989) 'Levels of Selection: an Alternative to Individualism in Biology and the Human Sciences', *Social Networks* 11, 257–72.

―― (1990) 'Weak Altruism, Strong Group Selection', *Oikos* 59, 1, 135–40.

―― (1997) 'Altruism and Organism: Disentangling the Themes of Multi-level Selection Theory', *American Naturalist*, 150, supp. vol., S122–S134.

―― (2002) *Darwin's Cathedral: Evolution, Religion and the Nature of Society*, Chicago: University of Chicago Press.

―― and Colwell, R. K. (1981) 'Evolution of Sex Ratio in Structured Demes', *Evolution* 35, 882–97.

―― and Sober, E. (1994) 'Reintroducing Group Selection to the Human Behavioral Sciences', *Behavioral and Brain Sciences* 17, 585–654.

Wilson, E. O. (1975) *Sociobiology: The New Synthesis*, Cambridge MA: Harvard University Press

Wilson, R. A. (2003) 'Pluralism, Entwinement and the Levels of Selection', *Philosophy of Science* 70, 531–52.

Wimsatt, W. (1980) 'Reductionist Research Strategies and their Biases in the Units of Selection Controversy', in T. Nickles (ed.) *Scientific Discovery: Case Studies*, Dordrecht: Reidel, 213–59.

Winter, W. de (1997) 'The Beanbag Genetics Controversy: Towards a Synthesis of Opposing Views', *Biology and Philosophy* 12, 149–84.

Wolf, J. B., Brodie, E. D., and Wade, M. J. (eds.) (2000) *Epistasis and the Evolutionary Process*, Oxford: Oxford University Press.

Wolpert, L. (1998) 'Comments on Epigenetic Inheritance in Evolution', *Journal of Evolutionary Biology* 11, 2, 239–40.

Wright, S. (1930) 'Review of *The Genetical Theory of Natural Selection* by R. A. Fisher', *Journal of Heredity* 21, 349–56.

―― (1931) 'Evolution in Mendelian Populations', *Genetics* 16, 97–159.

―― (1960) 'Genetics and 20th Century Darwinism', *American Journal of Genetics* 12, 365–72.

―― (1980) 'Genic and Organismic selection', *Evolution* 34, 825–43.

Wynne-Edwards, V. C. (1962) *Animal Dispersion in Relation to Social Behaviour*, Edinburgh: Oliver and Boyd.

Index

adaptation 10, 145, 171
  group 43, 79, 176, 186, 188–9, 196
additivity 79, 82–3, 87, 114–19,
  167–9
  and Wimsatt/Lloyd approach 79,
  114–19, 157
  see also epistasis
aggregate characters 48–9, 55, 57, 63,
  92, 99, 113–4, 119–20, 140
  see also: emergent characters
Alexander, R. D. 149
Allee, W. C. 175
allometry 81
altruism 11, 56–7, 66–8, 174–6, 191
  and group selection 177–8, 188
  and hypercycles 223–4, 230–1
  and kinship 182–4
  and the gene's eye view 144–5, 148
  reciprocal 180–5
  weak versus strong 192–7, 230
  see also cooperation; Hamilton's rule
altruistic behaviour, *see* altruism
apoptosis 238
Ariew, A. 33
Arnold, A. J. 19 n. 9, 54, 55, 56, 111,
  160, 204
Arnold, S. J. 25 n. 14, 81, 88
Aspi, J. 86 n.
assortative grouping 194–202
  see also positive assortment
avatars 210–11
averaging fallacy 137 n., 185, 189–92
Avital, E. 15, 149, 159
Axelrod, R. 180 n. 10, 181

Barrett, M. 126 n. 14
beanbag genetics 169
Beatty, J. 32
Beckermann, A. 49 n. 13, 106 n. 27
Berg, H. A. van den 157
bookkeeping objection 158–66,
  189–90, 225
  see also averaging fallacy; causality
Boorman, S. A. 179
Borgia, G. 149
Boyd, L. H. 86 n., 87 n. 11, 93

Boyd, R. 15, 159
Brandon, R. N. 18, 32, 38 n., 79,
  121–4, 158, 212 n. 7
Brodie, E. D. 115 n. 3
Burt, A. 145
Buss, L. W. 16, 17, 50, 53, 108, 127,
  129, 147, 148, 170, 218, 219,
  220–1, 225–9, 228

cancer 11, 238
Carson, S. 32
causal decomposition 25–31, 83–4,
  98–9
  and multi-level selection 66, 94,
  98–9, 199–202
  and random drift 31–3
  see also statistical decomposition
causality 113–4, 127–132, 201–2
  and correlation 76–84
  and multi-level selection 76–111,
  134
  and the bookkeeping argument 158,
  162–6
  non-realist views of 128–9
  see also causal decomposition;
    cross-level byproducts
causation, *see* causality
causation from below, *see* cross-level
  byproducts
character-fitness covariance 23–39,
  166, 190
  and causality 76–111
  and pluralism 135– 38
  and screening off 122–4
  at two levels 62–6
  species-level 207–8
  spurious 81–4, 90–3, 109, 129, 236
  see also Price's equation; cross-level
    byproducts
chicken-and-egg problem 43, 131–2
clades 212–17
Clayton, P. 49 n. 13
clonal plants 45–6, 161
Colwell, R. K. 196
conflict between levels 11, 149, 153,
  219, 226, 228, 229–30

conflict between levels (*cont.*)
  see also intra-genomic conflict, outlaw genes
context dependence 164–9, 172
  see also *epistasis*
contextual analysis 86–99, 118–121
  and diploid population genetics 154–7
  and neighbour approach 198–202
  and Price's equation 93–9, 136, 154–7
  see also cross-level byproducts
cooperation 181
  and evolutionary transitions 222–3, 228
  see also altruism, prisoner's dilemma
correlated characters 80–4
Crick, F. H. C. 151
Cronin, H. 159
cross-level byproducts 78–80
  and contextual analysis 86–93, 119–125
  and emergent character requirement 113
  and fitness decoupling 235–6
  and fortuitous group benefits 176
  and pluralism 129
  and reductionism 141–2
  and species selection 206–9
  in MLS1 84–99
  in MLS2 100–111
cultural inheritance 15, 158–60
cultural evolution 177
cytoplasmic outlaws 152–3

Damuth, J. 19, 25 n. 49, 54, 55, 56, 58, 59 n., 60, 66 n., 85–9, 92–3, 95, 114, 119–21, 140, 179, 203, 210–11, 229, 236, 239
Darwin, C. 10, 11, 15, 144, 174, 182, 183
Davies, P. 49 n. 13
Dawkins, R. 15, 16, 121 n., 129, 139, 143, 144, 146–7, 148, 149, 159, 167, 168, 170, 171, 180, 182 n., 191, 205 n., 211, 221, 222, 225, 226
Demeter, L. 222, 224, 230, 231
Dennett, D. C. 171, 173 n. 1
division of labour 43, 92, 223, 238
Dobzhansky, T. 203 n.
Doolittle, W. F. 151

downward causation 77, 92
  see also cross-level byproducts
drift, *see* random drift
Drosophila 152, 206
Dugatkin, L. A. 126, 127, 133
Dworkin, M. 130

ecological hierarchy 45–6
Edwards, A. W. F. 104 n.
effect macroevolution 207–8
  see also cross-level byproducts
Eigen, M. 223
Eldredge, N. 45, 46, 79, 104, 100, 141, 203, 204, 208, 213
Elster, J. 139
emergent character requirement 79, 106–7, 112–14, 119, 207–8
emergent characters 48–9, 55–7, 140, 207
  and contextual analysis 92, 99, 201–2
emergent relations 119–20
Endler, J. A. 25 n. 14, 77 n.
epigenetic inheritance, *see* non-genetic inheritance
epistasis 115, 167–9, 172
  see also context-dependence
Ereshefsky, M. 14 n.
evolutionary game theory 176, 181–5, 239
evolutionary stable strategy (ESS) 165
evolutionary transitions 16–8, 42, 159–60, 161, 178, 218–40
  and levels of selection 218–40
  and MLS1/MLS2 distinction 219, 229–33, 236–9
  genic versus hierarchical approaches to 225–9
  see also fitness decoupling
Ewens, W. 104 n.

Falconer, D. S. 36
Falk, R. 226
Feldman, M. W. 18, 43, 175 n. 5, 179, 191 n., 193 n., 194 n.
female-biased sex ratios 151, 196–7
Fisher, R. A. 13 n., 52, 104, 116 n. 4, 143–4, 167, 169, 175, 203 n., 211
fitness 10, 13, 19, 46
  allelic 53, 53 n., 68, 150 n., 163
  cell 101, 103, 231, 233–5, 238
  clade 212–17

Index

collective 53-6, 117, 122-4, 229-30, 237-8
frequency-dependence of 58, 104, 165
  genic 149-50, 154-5, 163-5
  group 10, 117-8, 156, 176, 196, 216, 232
  inclusive 125, 145, 148
  individual 191-3, 198-202, 216, 232
  organismic 101, 103, 144, 152, 208, 231, 233-5
  particle 53-6, 118, 122-4, 237-8
propensity interpretation of 32
realized versus expected 32-3
species 55, 206-19
fitness decoupling 79, 232-6, 238
see also *cross-level byproducts*
Fletcher, J. A. 195 n.
Flohr, H. 49 n. 13, 106 n. 27
focal level, see focal units
focal units, 56-9, 160-2, 178-9
Fontana, W. 17
Frank, S. A. 12, 17, 19, 20, 24, 26, 27, 30 n. 20-1, 64 n. 23, 68, 71, 104 n., 108, 144 n. 2, 145, 150
Fristrup, K. 19 n. 19, 54, 55, 56, 111, 160, 204
functional organization 42-3, 140, 205
fundamental theorem of natural selection 13 n., 104

Galton, F. 30 n. 19
genealogical hierarchy 45-6, 212
gene's eye view of evolution 142-72, 226-9
  and context-dependence 149, 166-9
  and kin selection 148-9
  and outlaw genes 149-53
  versus genic selection 146-9
gene's eye perspective, see gene's eye view of evolution
genetic conflicts, see intra-genomic conflict
genetic information 18, 221
genets 45-6, 161
see also clonal plants, ramets
germ line sequestration 48, 219, 220, 226
Getty, T. 86 n. 9
Ghiselin, M. 191, 205, 212

Gilinsky, N. L. 56, 208, 209
Godfrey-Smith, P. 15, 43 n., 52, 53 n., 69 n. 27, 116, 117 n., 126 n. 13-14, 133-9, 150 n., 157 n., 193 n., 194 n., 221
Goodnight, C. J. 68, 86 n., 95, 96 n., 97
Gould, S. J. 12, 15, 79, 110-11, 120-1, 139-40, 148, 149, 158, 167, 171, 203-5, 208, 209
Grafen, A. 19, 33, 69 n. 28, 85, 104 n., 144 n. 2, 152, 177 n., 182, 194 n.
Grantham, T. A. 114, 141, 207, 208, 217
Griesemer, J. 14, 16, 17, 219, 221, 224
Griffiths, P. 130-3, 137-8, 165

Haber, M. 216 n.
Haig, D. 151, 152
Haldane, J. B. S. 169, 175, 191 n.
Hamilton, A. 216 n.
Hamilton, W. D. 19, 62, 65, 66, 71, 86, 93, 117 n., 144-5, 151, 152, 176, 177, 180, 181, 182, 183
Hamilton's rule 68, 144 n. 2, 180
  see also altruism
Hammerstein, P. 12, 180 n. 9
Hansen, T. A. 208
Hecht, M. K. 206
Heinrich, J. 108, 159, 160 n.
Heisler, I. L. 19, 25 n., 49, 54, 55, 56, 58, 59 n., 60, 66 n., 85-9, 92-3, 95, 114, 119-121, 140, 179, 229, 236, 239
Hennig, W. 212 n. 7
heredity, see heritability
heritability 13, 213
  and mode of reproduction 186, 188
  and transmission bias 34-9
  collective 59-62, 71-4
  group 185-9
  in MLS1 60-2, 71-4
  in MLS2 60-2, 75
  particle 59-62, 71-4
heterosis 157, 167-8
heterosis argument 162-6, 190
heterozygote superiority, see heterosis
Heywood, J. S. 30 n. 20, 35
hierarchical organization 10, 40-6, 219-25, 227-8
  indeterminacy of 130-2, 138, 184-5, 190-1, 224, 239

hierarchical organization (*cont.*)
  interactionist conception of 41–6, 98, 132, 183
  nested versus overlapping 43–5, 99
  see also genealogical hierarchy; ecological hierarchy
  hierarchical structure, *see* hierarchical organization
Hodge, M. J. S. 175 n. 5
Hoekstra, R. F. 151, 152
Hoffman, A. 206
holism 93, 172, 224–5
  see also reductionism
Hull, D. L. 15, 16, 79, 205, 211, 212
Hume, D. 128 n.
Hurst, G. D. 145, 148, 149
Hurst, L. D. 145, 148, 149, 151, 152
hypercycles 222, 223–4, 230–31

individuality 101, 130, 218–20, 233–6
  see also evolutionary transitions
insect colonies 139, 161, 173, 174, 221, 222
interactors 146–7, 211
intra-genomic conflict 12, 149–53
  see also outlaw genes
Irzik, G. 82 n. 7, 122 n. 9
Iversen, G. R. 86 n., 87 n. 11, 93

Jablonka, E. 149, 159, 160 n.
Jablonski, D. 57, 60, 206, 208
Janzen, D. H. 46

Kalisz, S. 25 n. 14
Keller, L. 12, 18, 19, 147 n. 3, 219
Kerr, B. 34 n. 23, 52, 53 n., 69 n. 27, 117 n., 133–9, 150 n., 193 n., 194 n.
Kilmer, W. C. 179
Kim, J. 49 n. 13, 106, 129 n.
Kitcher, P. 126, 128–9, 165, 170
Krause, J. 55

Lack, D. 126
Laland, K. N. 18
Lamb, M. 149
Lande, R. 25 n. 14, 81, 88
Leigh, E. G. Jr. 153
levels of selection question
  synchronic versus diachronic formulations of 16–8, 219–25, 229, 236, 239–40
  see also evolutionary transitions
Levin, B. R. 179
Levins, R. 179
Levitt, P. R. 179
Lewis, D. 30 n. 18
Lewontin, R. C. 13, 15, 16, 113 n., 116, 134, 149, 158, 162–7, 173 n. 1, 211, 217
Lewontin's conditions 13–14, 210–11
  insufficiency of 37, 8
  violation of 96–9
  and Price's equation 34–9
Lieberman, B. S. 206, 208
life cycles 49–53, 160–2
linearity, *see* additivity
Lloyd, E. A. 79, 114–16, 120–1, 157
Loewer, B. 106 n. 26, 129 n.
Lorenz, K. 175, 181
Lukes, S. 139
Lyttle, T. W. 151

Mach, E. 128 n.
macroevolution 203–17
McShea, D. W. 40–6
major transitions, *see* evolutionary transitions
Margulis, L. 178
Matthen, M. 33
Maynard Smith, J. 12, 13, 15, 16, 18 n., 42, 43, 58, 101, 112, 130, 133, 147, 149, 151, 152, 161, 165, 173 n. 1, 176, 178–80, 181, 182–9, 205 n. 4, 207, 218, 219, 220, 221, 222, 223 n., 226, 228, 230, 231, 240 n.
Mayo, D. G. 56
Mayr, E. 121, 158, 169, 172
meiotic drive 68–9, 152, 154–6, 165–6, 190
  see also segregation-distorters; outlaw genes
Mendelian segregation 70, 150, 220, 223
  see also meiotic drive
Meyer, E. 82 n. 7, 122 n. 9
Michod, R. E. 12, 16, 17, 19, 25 n. 13, 42, 49 n. 14, 55, 70, 79, 101, 108, 116, 141, 144 n. 2, 145, 149, 150, 160 n., 178, 212 n. 8, 218, 219,

221–2, 223, 224, 228–9, 230–6, 237, 238
Mills, S. K. 32
Mishler, B. D. 212 n. 7
mitochondria 50
  uniparental inheritance of 152–3
monophyly 45 n. 9, 213, 217
Monro, K. 46
multi-cellularity 55, 101, 130, 159–60, 219, 221–2, 226
  Michod's model for the evolution of 55, 101, 219, 228–9, 231–6
multi-level selection 1 (MLS1) 56–62, 84–5, 117–8, 166
  and clade selection 216
  and group selection 178–80, 187–9
  and pluralism 136–40
  contextual approach to 86–99, 124 n., 136, 154–7, 197–202
  neighbour approach to 197–202
  Price approach to 65–71, 85–6, 124 n., 136, 150, 154–7, 195–7
multi-level selection 2 (MLS2) 56–62, 96–7, 100–11, 120, 178–80
  and pluralism 136–40
  Price approach to 74–5
multiple regression 81–4, 87–9, 198–202
mutation test 192–4, 196

Nedelcu, A. 42, 219, 232
neighbours 193–5, 198–202
Nelson, G. J. 213
neo-group selectionist revival 176–8
non-genetic inheritance 149, 158–60, 162
  see also cultural inheritance
non-random group formation, *see* assortative grouping
Northcott, R. 27
Nunney, L. 43, 85, 93, 107, 112, 173 n. 1, 191, 192, 194 n., 195–8, 200, 202, 212, 236

Odling-Smee, F. J. 18
Orgel, L. E. 151
outlaw genes 145, 147–53

Pan, J. J. 46 n. 10
part/whole structure, *see* hierarchical organization
Pearson, K. 81, 128 n.

Pedersen, B. 46, 86 n.
Platnick, N. 213
pluralism about levels of selection 125–139, 184–5, 224, 226
  and causality 128–30
  and hierarchical organization 130–3
  and multiple representations 133–9
  and the gene's-eye view 129, 169–70
Pomiankowski, A. 145, 151
Poore, A. G. 46
population genetics 115, 162–9, 190–1, 228
  and multi-level selection 68–70, 150, 154–7, 197
population structure 56–8, 98, 183, 194, 223, 230
positive assortment 68, 182–5, 224, 230
Price, G. R. 18, 24, 62, 65–6, 70, 86, 93, 135, 176, 181
Price, H. 128 n.
Price, J. S. 46 n. 10
Price's equation 12, 18–39, 228, 235–6
  and causality 25–31
  and group selection 19 n. 8, 62, 66–8, 85–6
  and Lewontin's conditions 34–9
  and multi-level selection 62–71, 74–5, 85–9
  versus contextual analysis 93–9, 154–7
prisoner's dilemma 181, 183–5
punctuated equilibrium 204–5

Queller, D. C. 34 n. 23, 35, 36, 68, 72, 144 n. 2, 173 n. 2, 222, 223

ramets 45, 161
  see also genets, clonal plants
random drift 31–3
Raup, D. M. 206
realism about levels of selection 125–139, 169–70, 224
  see also pluralism about levels of selection
recombination 152
reductionism 93, 139–142, 224–5
  and cross-level byproducts 141–2, 170–1

reductionism (cont.)
  and part-whole structure 139–40, 171
  and the gene's-eye view 142, 160–72
  versus holism 93, 139–40
Reeve, H. K. 18, 126–7, 133, 147 n. 3, 219
Reichenbach, H. 121
replicators 16, 145, 221, 222, 223
replicator-interactor conception of evolution 15, 16, 146–7, 210–11
reproduction 14, 46, 212–15
  collective 50–2, 58, 72, 236
  group 178–9
  sexual 151, 153, 168, 186, 207, 209, 217
Rice, S. 12, 19, 24, 25 n. 14, 30 n. 20, 32, 33, 34 n. 23, 36, 54, 96
Richerson, P. 15, 159
Ridley, M. 182 n.
Roughgarden, J. 34 n. 24
Roux, W. 225
Roze, D. 55, 79, 101, 219, 228, 231, 233–6
Russell, B. 128 n.
Ruxton, G. D. 55

Salmon, W. 121
Salt, G. W. 48 n. 13
Sapienza, C. 151
Sarkar, S. 116, 157 n., 171, 172, 226
Schuster, P. 223
screening off 79, 82 n. 7, 121–5
  see also cross-level byproducts
Seger, J. 19 n. 9, 69 n. 28
Segerstråle, U. 12 n. 2, 19 n. 8
segregation-distorters 66, 70, 145, 150, 151
  see also meiotic drive, outlaw genes
selection
  avatar 210–12
  cellular 11, 228, 231, 233–6
  clade 212–17
  collective-level 56–75, 84–111, 112–14, 115–19, 131–2, 154–7, 190
  direct versus indirect 25, 77–84, 90–3
  individual 66–8, 148, 173, 189, 191–8
  genic 52, 69–70, 115, 129, 142, 146–57, 165–6, 169–172, 189–90
  group 11, 43, 55, 85–6, 97, 155–6, 159, 173–202, 222
  kin 148, 176, 178, 180, 182, 183–5, 230, 239
  multi-level 12, 46–7, 56–111, 116–121, 134–9, 153, 183–5
  'of' versus 'for' 25
  on correlated characters 80–4
  organismic 115, 154–7, 165–6, 220, 228, 231, 233–6
  particle-level 50, 56–75, 84–111, 154–7, 190
  response to 36–8, 49, 60, 71–3
  soft 95–6, 118
  somatic 50, 53, 75
  species 55, 57, 100–1, 104, 110, 140, 203–17, 239
  stabilizing 25
  trait-group 66–8, 125, 177, 185–9, 230, 239
selection differential 23, 71–3
selfish genes 144, 148
  see also outlaw genes; gene's eye view of evolution
selfish genetic elements (SGEs), see outlaw genes
Shapiro, J. A. 130
shifting balance theory 115, 175
Skyrms, B. 173 n. 1, 181, 195 n.
slime moulds 51–2
Sober, E. 12, 19, 27, 28, 32, 41, 43, 54, 56, 66, 85–8, 97, 108, 113, 116–18, 121–5, 134, 137 n., 147, 149, 158, 160, 162–7, 173, 177–8, 180 n. 8, 181–9, 191, 192, 196–7, 217, 236
Sober's height example 85–7, 90, 235
social bacteria 130
sociobiology 223
'species are individuals' thesis 205, 212
Stanley, S. M. 203, 208
statistical decomposition 25–31, 66, 83–4, 94, 98–9, 199–202
Steen, W. J. van der 157
Sterelny, K 18 n., 42 n., 43, 44 n. 5, 101, 126, 129–33, 137–9, 141, 165, 170, 209, 211, 212, 213
Stevens, L. 97
Stidd, B. M. 207
superorganisms 222

supervenience argument 105–7, 114, 124, 129
suppression of competition 42, 150, 153, 222
Szathmáry, E. 12, 15, 16, 42, 130, 149, 151, 152, 161, 178, 218, 219, 221, 222, 223 n., 224, 226, 228, 230, 231

tit-for-tat strategy 181, 184
trait groups 43, 44 n. 5, 50, 56–7, 61, 66–8, 177, 183, 185–9, 230, 239
transmission bias 20, 26–31, 67, 70–1, 75
transposons 145, 151
see also outlaw genes

units of selection 13–14, 18, 145
versus levels of selection 18
units of evolution 14, 185–6
Trivers, R. L. 145, 176, 180, 181, 191
Tsuji, K. 19, 86 n.
Tuomi, J. 46, 86 n.
Uyenoyama, M. 43, 175 n. 5, 179, 191 n., 194 n.

Valen, L. van 213
vehicles 15, 221
Vermeij, G. J. 212, 213
Vrba, E. 58, 79, 100, 104, 106, 110–11, 112, 115 n. 2, 139, 141, 206–8, 212 n. 8, 217
Vries, H. de 203 n.
Vuorisalo, T. 46

Waddington, C. H. 240
Wade, D. L. 207
Wade, M. 19 n. 9, 25 n. 14, 51, 64 n. 23, 97, 115 n. 3, 141, 148, 175 n. 5, 177, 179
Walton, D. 121
Waters, C. K. 126 n. 13
Weismann, A. 12, 225
Werren, J. 145, 148, 149
West-Eberhard, M. J. 191 n.
Williams, G. C. 43, 79, 113, 121 n., 139, 141, 143, 145–6, 163, 168, 171, 176, 195, 205, 207 n., 212, 213, 225, 226
Wilson, D. S. 12, 19, 41, 43, 44 n. 5, 50, 52, 56, 61, 66, 96–7, 113, 116–18, 122–5, 134 n. 20, 137 n., 141, 148, 159, 165, 173, 177–8, 180 n. 8, 181–9, 191, 192, 194–5, 196–7, 230
Wilson, E. O. 221
Wilson, R. A. 133, 159
Wimsatt, W. 79, 114–16, 119, 157, 167
Winter, W. de 144, 169 n.
Wolf, J. B. 115 n. 3
Wolpert, L. 17, 42, 130
Wright, S. 115, 116 n. 4, 141, 144, 149, 157, 158, 167, 169, 175, 177, 204, 205
Wright's rule 204–5
Wynne-Edwards, V. C. 126, 146, 175, 178, 185

Zwick, M. 195 n.